Einstein's Beetle

MARK SOUTHWORTH

NEWTON'S LAWS OF MOTION AND GRAVITY, AND EINSTEIN'S THEORY OF RELATIVITY

AN EXPLANATION FOR PEASANTS BY A PEASANT
WHO UNDERSTANDS

I

'All physical theories, their mathematical expressions notwithstanding, ought to lend themselves to so simple a description that even a child could understand them.'

Albert Einstein

Preface

Where are you on your journey to understanding the universe?

I started mine, 28 years ago, with the casual aim of understanding Einstein's theory of relativity and no idea of the mountain I would have to climb, the wonders I would uncover, or the direction it would ultimately take me. Like millions of people, Stephen Hawking's *A Brief History of Time* sent me on my way and then confused the hell out of me, but in spite of this – or because of it – popular science book seemed to follow popular science book and a mild interest grew into a passionate hobby.

There was no plan to my reading. Some books were dedicated to Einstein's theories; others included broader aspects of cosmology and quantum mechanics. I didn't pore over explanations, reread difficult sections or become frustrated if I failed to understand. I simply found interesting titles and let the facts and explanations sink in and connect as they may.

For years I couldn't find answers to my questions and the big ideas escaped me. But when the leaping-out-of-the-bath insights did arrive, and the opaque became clear in a blink of understanding, rather than *Eureka*, more often than not I'd find myself exclaiming, *why wasn't it explained this way in the first place!* After toying with a lecture for interested family and friends – and with the short-lived Stuntman's Quantum, Big Bang Universe and Light club, more aptly known by its acronym SQUABBUL – I decided to follow bestselling author Maggie Stiefvater's advice: *Write the book you've always wanted to read but can't find on the shelf.*

Everything about Einstein's Beetle is designed to explain the physics, with the fiction playing the central role. Newton's laws are best explained in a moving car, to which everyone can relate, and all explanations of relativity ask that you imagine you are in a free-falling lift or rocketing a spaceship near the speed of light. By combining these imaginings into a consistent story that mirrors the books real purpose, through Danny you will experience and discover much of the physics for yourself. The fiction also allows for a question-and-answer interplay between Danny and Bert throughout the book. And guided by Danny's own understanding – or lack of it – at each stage, the story setting will help you remember the physics and build a complete picture of the theory.

Although the fiction is an adventure in its own right, Einstein's Beetle is the mysterious story of our universe and the discovery of its physical laws, and as with all mysteries, the answers unravel over time, not all in one go. So, wherever you are on your journey, don't panic about your understanding of the physics. Relax, let Danny be your guide, and enjoy the ride. Your understanding of the world is about to change.

Mark Southworth
London, January 2016

Contents

Einstein's Beetle

Part I

Newton's laws of motion and universal law of gravitation, and the mystery of the speed of light

old on add in the question and answer interplay, another
examine and without any doubt, you will understand no idea if
ally saw decide to

Bert's Universal Tour and Expedition

Summer Science Camp

Explore the mysteries of the cosmos and experience the cowboy life at The Missing Horse ranch, deep within the beautiful Grand Canyon National Park.

Join Bert for an inspirational one-week journey through the history of cosmology and the laws that have shaped our universe, covering key stages of the physics curriculum for years 9 -13.

Warning: The Expedition is **DANGEROUS** and not for lily-livered softies.

Log on to www.butae.com for full details and to register your interest/application. Parental consents required.

\-

uld meet
d be no
ar sign
ford is
ust fail
rmote a
in, now
m stars
ent.
or us to
rarity if
ated.
ransom
at ratio
ransient
cally pi
rnet, so
as given
ld?
ing of a
velation
dull and
ive up.
rating
in for
ied a
a chav
r mini
ll not tially

To th
Every m
in what
not tha
simply,
to find a
each lit
identica
never be
sometime
As fa
talk into
harbour
even if a
of mana
read a h
your fir
When
or perha
further, i
Of the
run insid
etched f
leads to
a foriegn
to assess
in so far
veering
inch awa
to achie
yet neit

everything. This idea is not of itself new, but it helps us build the clearest picture of

1

Danny shovelled sand from the roadside onto the uppers of his left boot, placed his right boot on top, and with his knees bent and arms wide, started grinding it into the leather. As his leg shuffled back and forth, his head flicked from side to side and his bottom lip grabbed his top one for a hug.

The road sliced open the plain's thorny brush skin from the mouth of the red sandstone valley, half-a-mile to his left, to the spirit level horizon on his right. It sharpened to a point and rolled out of sight, as though it might circle the globe, but only the sun had emerged from the valley and for an hour none of its rays had come bouncing back.

Danny dropped his head to his chest and swapped boots, using the toe of the left to scoop sand onto the right before balancing back on top, like a one-legged pigeon coming in to land.

He was supposed to be dressed as the Rango Kid – outlaw, gunslinger, winner of six duelling gold medals, and his avatar in a virtual reality game called Wild West World. But his tan boots looked too shiny, his red-checked shirt too baggy, the crotch of his jeans too low and he'd lost his Stetson.

The outfit was Condition Four in the deal he made with his mum after failing his end of year physics exam. She agreed in the end provided it was the last condition, which he argued was itself Condition Five and didn't make sense. She reminded him that Condition One meant she could refuse whatever she liked, and making her spit on her hand – she refused to do it with blood – he quickly shook it to seal the deal.

Condition One said that if he didn't attend science camp on their summer holiday to visit Great Uncle Gil in Arizona, his mum would sell his Xbox.

Danny stepped back onto both feet and inspected his boots. He was about to start working their sides along the edge of the tarmac when a grey shape emerged from the shadow of the valley and flared into a bright orange blob.

Convinced it was a bus-full of his fellow campers, and expecting it to arrive in no time, he dived for his rucksack and held out his arm. But the blob spent a minute growing into a bigger blob, and he wasn't sure what for another two. At the end of the fourth, it finally resolved into an old Volkswagen Beetle with a large overloaded roof rack, a small sausage-shaped rear window, and a Ⓓ sticker in the middle of its boot.

The fact that the Beetle was in reverse struggled to compute with Danny, as it cruised up to him without a hint of slowing down. As it sailed silently past, the fact that its sole occupant – an old man in a khaki shirt and matching shorts – lay snoring beneath a newspaper along the backseat didn't exactly help. It was BERT'S UNIVERSAL TOUR AND EXPEDITION in large blue letters on a banner across the front of the roof rack that kick-started his brain, and with a 'Whoa. No. WAIT!' to engage first gear, he started sprinting after the car.

'Hey! Mister!' he called, drawing alongside the open passenger window. 'You're moving!'

The old man jolted the newspaper off his head to reveal a dandelion puffball of thin, snow-white hair. A stained wedge of moustache sat beneath a bulbous nose, and mottled skin had flowed and set, like cooling lava, in great furrows across his brow and swinging pouches beside his chin. To Danny, he looked just like the photograph of Albert Einstein that used to hang in his dad's study – and was now hidden away, along with everything else, in boxes in the basement. But though the old man's face suggested he was pushing ninety, magnified through thick, rimless spectacles something about his pale-blue eyes didn't.

'Oh, hello,' he said, smiling up at the skinny, freckled kid with wide, brown eyes, bouncing around the Beetle's passenger window frame. 'Can I help you?'

'Stop the car … it's me,' panted Danny. 'I saw the banner.'

The old man cupped a hand behind an earful of pink hearing aid.

'What was that?'

As well as being deaf, he sounded German, using *z*'s for *th*'s and pronouncing his *w*'s like *v*'s. 'I'M DANNY,' shouted Danny.

'Danny!' cried the old man, dropping his jaw into an open-mouthed grin. 'You have arrived. Ha, ha! Morris told me you would be coming this way. Welcome to my Universal Tour and Expedition.'

'Bert?'

'That's right. And look at you, jumping with excitement. Cannot wait to get going, eh?'

That threw Danny and he disappeared below the window frame, clawing at the side of the car to recover his stride.

'Me too," continued Bert, scooping a clipboard from the Beetle's floor. 'I hardly slept last night.'

'I'm trying to keep up with you,' spluttered Danny, popping back into view.

'Ah yes, here we are. Daniel P. Robinson, aged 14, all the way from London. I have been to London.' Bert shoved a hand deep into the pocket of his shorts – at least Danny thought they were shorts; thin, white legs emerged so briefly before disappearing into old hobnailed boots, it was difficult to tell. 'Jump in,' he said. 'We'll soon be off.'

'You're already off!' cried Danny. 'You're in reverse!' Something hard in his rucksack kept bouncing against his back and the crotch of his jeans was beginning to chafe.

'We will not get anywhere without … this.' Bert pulled out a cherry lollipop, holding it up triumphantly until he noticed what it was. 'That's not it.' And his arm disappeared into his other pocket.

'JUST STOP THE CAR YOU –'

Danny's eyes popped wide as he started to lose ground.

Bert's narrowed as they watched him reverse into view in the windscreen and drag himself backwards down the side of the bonnet. 'Are you alright?'

'Alright!'

'Is it asthma?'

Danny's mouth fell open, but he clenched it shut. 'You'll crash,' he shouted, and exploding his last reserves of energy in a desperate belly-flop dive, he landed diagonally across the bonnet with his face pressed into the glass.

'Good thinking,' said Bert. 'Put your rucksack on the roof rack.'

Danny glowered through the windscreen, but seeing that the old man had returned his attention to his shorts, he grabbed hold of the metal rail above his head and clambered to his knees.

The roof rack supported a mountain of folded furniture and wooden crates, capped with a white, metal chest and secured with heavy-duty ratchet straps. Assorted aerials and satellite dishes sat around its top rail, with two metal dustbins on their sides pointing back off its rear corners, like rocket thrusters.

Slipping off his rucksack and wedging it under one of the straps, Danny surveyed the rest of the battered Beetle. Bits of twig and tufts of grass seemed to sprout from any and every gap, forming hedges around fields of dented body panels and fringes along the crash-bars wrapped around its front. To top it off, the entire body was coated in a thick crust of dried mud and rusty sand, as though it had been sprayed on in a paint shop.

It occurred to Danny that the Beetle was in almost as bad condition as its owner, and as the owner appeared to have disappeared, he reached around the corner of the roof rack, searched his right boot down to the narrow running board beneath the passenger door, and like the Rango Kid robbing a stagecoach, swung his body up against the side of the car.

Bert was on his hands and knees on the floor, feeling under the seats.

'Will you get up and look where you're going. What are you looking for anyway?'

'The ignition key.'

Danny's eyes snapped to the steering column and the key buried in its side. 'It's in the ignition. And it would be because we're reversing up the road. Right now!'

'Look at the speedometer,' said Bert, his face appearing in the gap between the seats.

Danny shuffled along the running board towards the back of the Beetle and peered through the glass. In the middle of the dashboard an oval, green screen displayed '**0 mph**' in large black letters. 'Well of course it says that; speedos don't work in reverse. Anyway, can't you tell we're moving?'

'No, but I am pleased that you have stopped.'

'It might not look like I'm moving to you –' Danny glanced down to his boots and the grey tarmac running smooth beneath their heels, 'but we are to the ground.'

'Ah, so you understand.'

'I understand you ain't got five beans in the wheel,' muttered Danny. He'd discovered that he could cuss or mock people – and get away with it – by playing around with old western slang he'd picked up from other gamers at Wild West World. Most of the time he didn't know what he was saying either.

Bert hauled himself back onto the rear seat. "Oo look!" he said, bolting upright. 'Everything's moving.'

'Y-e-s,' nodded Danny, as though to his three-year-old sister, Alice. But Bert was smiling with wonder as though he too should be in a booster seat. 'So? Aren't you going to stop the car?'

'Hu? Oh yes. Wait there, my boy. Wait there.'

Bert squeezed between the front seats, dropped in behind the steering wheel and swung open the passenger door. Reaching out, he grabbed a startled Danny by the shirt and dragged him headfirst into the Beetle.

*

It was only a simple newspaper ad, buried in the pile of glossy brochures his mum had been secretly collecting, but Danny had fought hard for Condition Two of the deal – the choice of science camp – and Bert's Universal Tour and Expedition seemed the best bet at the time.

One night after dinner, using Alice to block him in, she dropped the pile onto the kitchen table and put him under covert surveillance by pretending to clean the hob. After yawning through the two- to three-week courses offered by the first few camps, the word *DANGEROUS* leapt out and made him sit upright. A beat after finding *cowboy life*, *Grand Canyon*, *one-week* and *ranch*, he pinned the advert to the table with his finger and announced, 'I'll go to this one.'

It started well, even if he did lose his hat.

They spent a night at a local hotel and then his mum dumped him back at Phoenix Airport and into the clutches of Condition Three – Linda, his dedicated airline chaperone for the early morning flight to Flagstaff. She wouldn't shut up and kept calling him Daniel, until he lost her in the departure lounge by ducking behind a baggage car and diving into a shop. Half-an-hour later, all puffing cheeks and flapping hands, she collared him as he swaggered up to the gate in a pair of hand-tooled leather boots and a casually tilted Stetson. But after informing him that his scheduled lift couldn't make it and new arrangements had been made, she left him to his iPod for the rest of the flight.

The new arrangements turned out to be Earl, a conspicuously real cowboy, even in a full-face racing helmet; a buttercup-yellow Kawasaki Z1000 motorbike; and a heart pumping, forty-minute Grand Prix to a screeching stop in the middle of nowhere. Danny was still grinning as he handed back his helmet. By the time he realised he should have stuffed his Stetson in his rucksack instead of tying it by its chinstrap, Earl had popped a wheelie and roared off up the road. An hour later he was scrambling off an old man's bony lap wishing he'd given a lot more thought to Condition Two.

'I am so pleased you made it,' said Bert, vigorously pumping Danny's twisted arm. 'I was worried when Morris told me he would be unable to pick you up. I've had a lift with Earl on that contraption and it scared the willies out of me.'

'Why didn't you just stop the car?' said Danny, pulling his hand free and checking the rear window. But the Beetle was still in the middle of its lane with no other traffic in sight.

'I told you,' said Bert. 'We are not moving. Look at the speedometer.'

Danny smiled, but the old man's watery-blue eyes looked back through his glasses, like puffer fish floating innocently in goldfish bowls. 'Never mind the speedo, look at the bushes and everything moving away!'

'Exactly.'

'What do you mean, exactly?'

'Look at the bushes and everything moving away.'

Danny looked to the windscreen, but he already knew what Bert meant. He jumped back to the rear window, but now that didn't look right either. An ocean of thorny scrub appeared to be drifting towards the back of the Beetle, as though it was being projected onto a screen. He started, his eyes swivelling from side to side. Something else was wrong. There was not a breath of sound from the engine, or outside, and he could feel no vibrations or bumps whatsoever. 'What sort of car is this?'

The Beetle looked like it had been built in a World War II Messerschmitt factory. Left-hand drive with only two pedals, a long automatic gearstick, and stripped of every original feature other than the steering wheel and the seats, it was all riveted metal, functional mechanics, latticed-steel roll cage and empty lollipop wrappers. There were no switches or handles to control the windows, or anything else, but a small keyboard sat beneath the oval speedometer and a trunk of multicoloured wires descended from the dashboard below it, branching out along the floor pan and beneath the doorsills, like thick roots searching for water.

There was also, Danny realised tilting back his head, the distinct odour of old socks.

'She is a 1954 Type One Volkswagen Beetle, but I have made a few modifications,' said Bert, his jaw dropping into another toothless grin. 'I have fitted a powerful electric motor, a floating hydraulic suspension system, full-body soundproofing and many other features; all computer controlled and …' he added, with a nudge to Danny's ribs, 'voice-activated. The screen, here, is the computer monitor with a manual interface below –'

As if on cue, a beep sounded and the '**0 mph**' on the screen started flashing.

'This is a highly sophisticated program so I was expecting a few problems with its teeth,' said Bert, his fingers dancing on the keypad. He gave the top of the dashboard a thump and received a beep in reply. 'Ah, you are right; we are moving. The speedometer must have been stuck.'

'We've been reversing the whole time,' said Danny, dragging his eyes from the rolling road. 'It only looks like everything else is moving because we're moving.'

'Oh no. Look. We are not moving backwards.'

The speedometer now read '**834 mph east**'.

Danny strangled a laugh. 'You don't believe that.'

'It is never wrong.'

'You just said it was stuck.'

'Yes, but it was still correct.'

Danny snorted and pointed at the windscreen. 'If we're doing 834 mph east, how come that valley's moving away as though we're reversing at 10 mph west?'

'Six.'

'Is that all?'

'You said yourself, it depends on your point of view.'

'That changes what you see, not what's actually happening.'

'How do you know what is happening? Have you ever sat on a train in the station when the train next to you starts to move and for a moment thought *you* were moving? Or your train pulls away but you think the other one is moving?'

'Yeah, but you get that feeling even if you're looking at the station.'

'Exactly!' beamed Bert. 'We have uncovered the first great mystery of our universe: *there is no such thing as standing still.*'

Danny's freckles bunched together as he started to laugh then parted almost as quickly. 'Are you doing all this deliberately?'

'For that first couple of seconds, sitting on a train, you cannot feel whether it is you who is moving or another train or the station, so why are you convinced that we are reversing west in my Beetle when you cannot tell that we are moving?'

'That's just your special suspension and sound proofing.'

'They only dampen the vibrations and sounds from the road and the engine.'

'I can see we're moving.'

'All you can see is *relative* movement. In other words, like the trains, something is moving but you are not sure what.'

Bert pulled a table tennis bat and ball from the pocket of his door and passed them to Danny. 'Start bouncing the ping-pong ball and try to keep it going.'

After a couple of attempts, Danny got the ball bouncing, more-or-less straight up and down, about 12 inches off the bat.

'What do you notice about the behaviour of the ball?'

'Nothing.'

'That's right. If my Beetle were large enough, we could have a game of table tennis right now and it would be identical to playing in your living room at home or a sports hall anywhere in the world. And this is no illusion or trick. Every physical process you associate with standing still at the roadside applies equally to us here in the car and so everything out there must be moving.'

'Roads and bushes don't move,' said Danny, catching the ball in mid air.

'Of course they do, along with everything else on the surface of our planet.'

Danny opened his mouth then closed it again.

'The Earth revolves once every twenty-four hours in a direction we call east. That is why we get night and day, and why the sun appears to rise in the East and set in the West.'

Danny had known for years that the Earth is continuously moving, but it had never occurred to him that standing still was the odd bit. 'We're standing still and moving at the same time.'

'Which means the idea of being completely motionless, that standing still is supposed to represent, does not really exist and everything must be moving.'

'Everything?'

'Everything. The Earth is rotating and orbiting the sun. The sun is one of billions of stars orbiting around the centre of our home galaxy, the Milky Way. The Milky Way is one of the 50,000 galaxies in a rotating band called the Virgo supercluster. And the Virgo supercluster is constantly separating from every other cluster due to the expanding spacetime of the universe itself. There is no special place or point called zero-miles-an-hour against which to measure an object's speed and position, and this lack of an Absolute Position means we have no idea where anything is heading or how fast it is getting there.'

'Then how come we can work out where everything is?'

'That's a good question: What can we actually measure about a moving object or body? But it is confusing being here on Earth, so let's leave our planet and imagine we are in a spaceship, all on our own, out deep in empty space.'

Suddenly four black screens rose silently up the outside of the Beetle's windows and squeezed out all the light.

2

How did you do that?' Danny smiled into the void of the Beetle's screened interior. It didn't occur to him that they were still reversing along the road at six mph heading west.

'Welcome to Spaceship Beetle,' said Bert, 'all alone in the depths of empty space.'

With his pupils expanding as fast as his imagination, Danny could see pinpoints of light emerging from the darkness around him; the faint twinkling of a thousand stars giving shape to the windows as they surrounded the Beetle in space.

'The window screens act like a flight simulator to form a computer-generated star field,' said Bert. 'And because the stars are so far away, we have absolutely no idea if we are moving or not.'

A beep sounded and '**0 mph**' flashed onto the computer screen, bathing the interior in its radioactive-green glow.

'If everything's moving,' said Danny, 'we must be moving.'

'Then what is our velocity? In other words, at what speed are we moving and in which direction?'

Danny looked out at the motionless stars and shrugged.

'We cannot measure anything about our motion and we call this total lack of information standing still. But what if another spaceship comes rocketing towards us?'

'What's that?' said Danny, pointing to the windscreen.

'It's another spaceship rocketing toward us.'

A speck of white light was growing fast in the middle of the black screen. Streaking through space, it separated into a rainbow of colour – and an oddly familiar shape – and shot past the driver's side of the car.

'Wow!' said Danny, spinning to watch its light dwindle in the rear window.

'That was Spaceship Beetle II, and now we can make a measurement. We can say that, relative to us at zero miles per hour in the Absolute Position, she passed by on the driver's side at …'

The computer beeped '**1000 mph**' onto its screen.

'But what happened from Spaceship Beetle II's point of view? A minute ago, her passengers were alone in empty space with no idea that *they* were moving. They made themselves the Absolute Position and –'

'We came flying past them.'

'According to their measurements, we passed on their driver's side at 1000 mph moving in the exact opposite direction. It could be up and down, east and west or each appearing to orbit the other, like the Earth and the sun. But if a body is moving one way relative to us, then we are moving at the same speed in the opposite direction relative to them.'

'And both points of view are right?'

'Exactly. They are equally real and we can only measure our velocities, or speed and direction, relative to each other.'

'What if there are other spaceships?'

'It does not matter how many there are or how they are moving, each will measure the velocity of every other spaceship relative to themselves as the Absolute Position.'

'That means everything's moving but everything thinks it's standing still.'

'And every point of view measures a different result, yet every measurement is correct because everything you see happen from that point of view is really happening. Let's bring in a third spaceship and see what we can measure.'

Danny blinked as the black screens descended, flooding the Beetle with crisp, morning sunlight. 'Where are the spaceships?'

'The first is Spaceship Earth,' said Bert, nodding to the ground spinning slowly beneath them in the windscreen. 'When you were standing on Spaceship Earth, even though you were revolving east with its surface, you measured our second spaceship, Spaceship Beetle, to be reversing past you at six mph heading west. From the point of view of Spaceship Earth that is exactly what happened, and it is how we are moving right now. But here on Spaceship Beetle we

22

can only measure that Spaceship Earth is moving past us at six mph heading east.'

'So the ground really is going past us?'

'As much as we are going past it. We are simply moving at the same speed in opposite directions from each of our points of view.'

'What about the 834 mph east?'

'That is from the point of view of our third spaceship, Spaceship North Pole, which is one end of the stationary axis around which the Earth spins. So if we are standing on Spaceship North Pole –'

'We can say we're the Absolute Position again.'

'Now at our current latitude relative to Spaceship North Pole, Spaceship Earth is moving with a velocity of 840 mph east and carrying with it Spaceship Beetle, which is moving west along its surface at six mph.'

'Oh-h, Spaceship Beetle's going six mph slower than Spaceship Earth,' said Danny, as the computer beeped '**834 mph east**' back onto its screen.

'And because velocity measurements include the *vector*, or direction of motion, we can add and subtract them to find the relative speed between any two bodies from any point of view. If two cars are moving in opposite directions relative to the Earth, one with a velocity of 40 mph east and the other 30 mph west, we can *add* their velocities together to find their speed relative to each other.'

'Seventy miles an hour.'

'That makes sense because each car measures itself to be standing still with the other either approaching or receding at their combined speed. How are the cars moving relative to the North Pole?'

Danny paused for a think. 'Spaceship – I mean the Earth's revolving at 840 mph east, and the first car's doing 40 mph in the same direction so its velocity is 880 mph east. The other car's doing 30 mph west, which is like moving 30 mph slower east, so it must be 810 mph east.'

'What do you notice? From the Earth's point of view the cars are moving in opposite directions, but they are moving in the same direction relative to the North Pole.'

'Oh yeah.'

'Now when we *subtract* their velocities, we get the same approach or separation speed.'

'880 minus 810, still 70 mph.'

'Each point of view measures a different pair of velocities that could have the cars moving in opposite directions or the same direction, but they always add or subtract to make the same approach or separation speed because that is how the bodies are moving relative to each other and that event must remain the same for everyone.'

'So we're not actually moving at different velocities all at the same time?'

'No, but we can be measured to be moving a million different ways from a million different points of view because *we cannot give a body an exact position in space*. When you were batting the ping-pong ball, from our point of view in my Beetle the ball bounced straight up and down in the *same space* as the ground moved east *through space* at six mph. To someone standing at the roadside watching us pass, they occupied the same space and it was the ball inside the Beetle that was moving through space; bouncing along the road at six mph in eight-foot jumps heading west. And from the ping-pong ball's point of view, my Beetle and everything in it bounced *down and up* on the spot, and the bystander and the road –'

'Bounced down and up at six mph in eight-foot jumps heading east,' finished Danny, nodding like a dashboard dog.

'In a universe in which everything is moving no single point of view is special and we can only make a relative velocity measurement from a specified point of view. Ha, ha!'

Bert started jigging on his seat, and grabbing Danny's hand he held it aloft, as though they were hoisting the FA Cup. 'That, my friend, is the *Principle of Relative Motion* and that simple concept has revolutionised our understanding of the universe. It was first discovered in the early seventeenth century when an Italian called Galileo Galilei used a ship, instead of a Beetle, as a stable platform for performing experiments on motion. In the right conditions ...'

But as the old man breezed on, Danny wasn't celebrating; he was staring at the windscreen wondering why they'd already started on the science. Without bothering to check the actual details, he'd convinced himself that Bert's Camp would be a school for cowboys more than geeks, with the odd science lecture and interactive experiment interrupting an otherwise full program of riding lessons, shooting competitions, and horseback tours of the Grand Canyon with his cohorts – a wild-bunch of Western-loving Dudes only there because their mums had threatened to sell their Xboxes ... So where were they? And why was Bert there with just him? Did he collect everyone this way, or did each teacher dress as a different scientist and pick up a few each? ... How far away was this Missing Horse –

'We're still in reverse!' Danny yelped, spinning to check the rear window and wondering why he bothered.

'Yes, it is probably time we measured our motion relative to the Earth,' said Bert, a pipe nodding in agreement at the corner of his mouth. 'I know. Let's switch off the engine and accelerate to a stop.' With a wave of his wiry eyebrows, he plucked the key from the ignition, put his hands behind his head and crossed his feet onto the dashboard above the computer screen.

Whoopee, thought Danny, but he couldn't help himself. 'If we roll to a stop we'll be slowing down not accelerating.'

'Acceleration and deceleration are the same event happening in opposite directions depending on your point of view. Look. Relative to the Earth we are slowing down, but relative to the North Pole we are accelerating.'

Bert pointed to the computer screen, and as the world started slowing in the windows, Danny watched the speedometer jump to 835, stroll to 836, and then crawl more and more slowly through 837, eight and nine.

'There,' said Bert, pushing a button on the keypad, as the planet rolled imperceptibly to a stop.

The computer displayed '**840 mph east**' with a small '**cc**' blinking regularly underneath.

'What's that?' said Danny.

'Cruise control.'

Danny started to laugh. Swaying open-mouthed in his seat, he looked from the speedometer to the stationary bushes lining the road then suddenly bobbed round to face Bert. 'Never mind dressing up as Einstein, you should be Professor Pat Pending.'

'Ah,' said Bert, grinning sheepishly and fiddling with his glasses.

'Yes, I am told the older I get, the more I look like Einstein when I was – I mean when he was my age.'

'How old are you?'

'Older than I look.'

Danny's head dipped into his shoulders as he blew another laugh out through his nostrils.

'It's the hair,' continued Bert. 'I have always had a full head and since it turned white, I admit I have let it grow to add to the effect.'

'Are you going to pretend to be German the whole time?'

'Oh no, this is my real accent. I am German by birth and did not emigrate to the United States until I was fifty-four. And you know what they say about mad dogs and old tricks.'

Danny did an impression of an interested smile, and Bert smiled back with as much conviction then looked away. 'But enough of me,' he said, bouncing back with a genuine grin. 'Do you see how much is becoming clear now that we understand the principle of relative motion? And this is just the beginning.'

That's what was worrying Danny, the beginning of what. 'Bert?'

'It is now 1609,' said Bert, poking his pipe into Danny's ribs as the computer beeped and '**1609**' appeared in large black numbers across its screen. 'Galileo is about to destroy another long-held belief in the way bodies move. Again, it is a belief we experience every day, yet it could not be further from the hidden truth.'

'What belief?'

'The Ancient Greek philosopher Aristotle's belief in a body's preferred state of rest.'

It sounded boring. 'Hey Bert, about your science camp. Where are the other –'

'And you believe it too.'

'How do you know?' said Danny, not sure whether he did or not. His preferred state of rest was lounging on his bed in front of his Xbox.

Bert started searching behind his seat. 'You did not seem surprised that we came to a stop when I switched off the engine?'

'No. Everything stops if it's not –' a smile crept onto Danny's face, like a clown turning up at a wake. 'You mean nothing really stops because there's no such thing as standing still.'

'Exactly,' said Bert, twisting back to the front and dropping a powder-blue crash helmet over Danny's head. 'Do that up.'

He pulled a second, yellow, open-face helmet from between the seats and filled it with fluffy white hair. 'Whether it is a ping-pong ball bouncing to the floor or a car when you switch off its engine, here on Earth every moving body comes to a stop and remains at rest unless a force acts to keep it moving.'

Danny peered down his nose to his fingers grappling to connect the straps beneath his chin. 'What's with the helmets?'

'The question is,' said Bert, clipping his chinstrap together and cinching it tight, 'what does standing still *really* mean?'

'It means you're still moving,' said Danny, settling for a granny knot.

Bert snapped home the buckle of his seat belt, as Danny searched for the end of his. 'But what is that motion and why is it special? Fortunately, we experience a huge clue as to how bodies want to move, every day.' He gave the roof of the car a thud. 'Especially in one of these.'

'Go on then, what?'

'INERTIA-A-A!' sang Bert, flooring the accelerator pedal.

'Whoa-agh!' cried Danny, as his seat-headrest knocked his helmet over his eyes; a force started squashing it against his nose; and the Beetle shot up the road like a Ferrari.

'Yeehaa!' whooped Bert. 'Can you feel it?'

The computer beeped and the force disappeared, giving Danny enough time to shout, 'I can't see a th-aagh!' before throwing him into the passenger door pillar as the Beetle turned sharply through ninety degrees and skittered across the road.

Bert let the steering wheel spin through his hands, returning the Beetle to a straight line and Danny to his seat to the sound of another beep. 'It's because we are not standing still any more,' he shouted, grinning wildly as they accelerated down a narrow track between dense brush walls, into the depths of the plain.

Glued to his seat, Danny winced inside his helmet as stiff branches ricocheted off the sides of the Beetle, like machine gun bullets. Another beep sounded, the force disappeared, and not sure whether to restore his sight or find something to grab, he shouted

'Wait!' with an arm trying each and 'Oof!' as he failed to make either.

Danny barged into Bert, head butting the side of his helmet as the Beetle made a second sharp turn, then pushing hard against him as she slid sideways through it.

The Beetle straightened, beeped, and charged towards a dense bush wall, forcing Danny to grab the frame of his seat and his instinct to let out a scream. He broke off as the force disappeared with a beep, but in another attempt to restore his sight, he let go of the seat-frame with both hands as Bert hit the brakes with both feet.

Danny piled into the windscreen, like a prop forward into a scrum, as the Beetle skidded out of the track and to a stop in the middle of a sandy clearing. He flopped back to his seat to another beep, but Bert powered the Beetle into a right-hand doughnut and they went round and round – Danny squashed against the door wailing 'Urrrghhh' and Bert secure in his seat belt singing 'Weeeeee'.

'Do you see?' said Bert, grinding the Beetle to a halt. 'Aristotle was wrong. A body's preferred state of motion is not to come to a stop.'

'Tell my head that.' Danny's stomach inflated his cheeks and the gas forced its way out through his lips, like a horse. 'I feel a bit sick.'

'Better still, let your head tell us. Look.'

Danny twisted his helmet until his face popped into view and found Bert pointing to a blue, powdery mark on the windscreen. 'What's that?'

'That chalk mark was made by your helmet when the car came to a stop but *you* did not. Look, there is a mark here.' Bert pointed to a second smudge on Danny's seat headrest. 'And here on the passenger door pillar, and here on the side of my helmet.'

Danny inspected his blue hands and looked up. 'Where are we?'

The Beetle had come to a stop in a circus-ring sandpit surrounded by an audience of tall sage stems and squat creosote bushes, and facing the only way in or out – the narrow track down which they'd charged.

'You tell me,' said Bert. 'How did we get here from the road?'

'Well –' Danny was about to fire back, how the heck should I know, when he realised he knew exactly, 'we … er … took off like a rocket then turned hard left, speeded up and turned right. You put your foot down again, rammed on the brakes, we came to a stop and went round and round. Oh and there was lots of beeping going on.'

'That's right,' said Bert, rocking Danny sideways with a punch to the shoulder. 'When a force acts on a body, it does not make it move; it makes it change the way it is already moving.'

'I didn't say that.'

'You said we accelerated, turned, accelerated, turned again, accelerated, decelerated, then turned round and round.'

'Oh yeah.'

'But when the forces that were making you change your speed or direction stopped acting ...'

'It felt like I was standing still.'

'Exactly! *The preferred state of motion of a freely moving body is to keep moving at the same speed and in the same direction as it is already moving* because that is precisely what you did until you were forced to do otherwise.'

Bert removed his helmet; his hair springing out, like a released feather duster. 'The first mark was here on your seat headrest. My Beetle wanted to remain standing still at 840 mph east, or zero mph relative to the ground, and you wanted to remain sitting in it at that same velocity. When the engine forced her to accelerate, you felt a force pushing you into your seat, but that was your seat pushing into your back, forcing you to accelerate against your natural inertia to remain at the same speed. As soon as the car stopped accelerating, you found yourself standing still again at ...'

'**40 mph east**' beeped onto the screen.

'The computer beeped every time the forces stopped acting?'

'That's right, and we found ourselves in uniform motion again, standing still at a new constant velocity.'

Bert pointed to the smudge on the passenger door pillar and to the side of his own helmet. 'You made these side marks when friction between my Beetle's tyres and the road forced her to turn, first 90 degrees to head north then 90 degrees to head east again. But you continued moving in the direction she was heading before making each turn; until the passenger door and my body forced you to turn against your natural inertia to keep moving in the same direction. Finally, we accelerated to stand still at 40 mph heading east again, and you made the last chalk mark when I applied the brakes, friction forced my Beetle to slow down, and you carried on moving at that same velocity until you –'

'Hit the windscreen, tell me about it.'

'We call this natural motion at a constant velocity, *inertia*, and as you demonstrated, a body will remain standing still in the same inertial state or *frame of reference* until acted on by an outside force.'

'What about when we were reversing up the road?'

'We cheated a little. We used one outside force, the engine, to keep us going at a constant six mph and cancel out the other outside force, friction with the atmosphere and the ground, that was trying to slow us down. But if forces are acting, they will change your inertial frame and that is why you should always wear your seatbelt.'

'Yeah, when did you put on yours? And why didn't you just explain it in one of the lectures or something?'

'Minds are changed through observation, not through argument. I wanted you to see for yourself.'

Danny was trying to think of something else to moan about when a horrible thought rose from his subconscious, like a fart bubble popping out of bath water. 'Bert? Where are the other students?'

'What other students?'

Danny's eyeballs bulged as the realisation hit them from behind. 'It's just you and me!'

'You, me and Morris. What made you think there would be others?'

'Oh-h, I don't know,' said Danny, flopping his arms up and down. 'Who's Morris?'

'Maurice is a proud Frenchman of aristocratic birth, or so he claims. I started pronouncing it Morris to wind his clock and it stuck. He is a retired physics teacher as well, though he has many years on me, but for the past five he has been my ranch manager, housekeeper, gardener and friend. He will be helping us on our expedition, doing the behind-the-scenes work. He was supposed to collect you from the airport this morning but his car was stolen last night.'

Danny dropped his helmet to his hands and shook it slowly from inside. 'Science camps always have loads of students.'

'I know. Aren't you lucky? Most are not real camps at all; they are schools with swimming pools. Our expedition is special. We are not just scientists, we are time travellers.'

'You mean when your computer jumps to a new date.'

'More than that,' beamed Bert, with enough enthusiasm for both of them. 'We are going to tour through the history of cosmology on the greatest journey humankind has ever made, exploring and solving the mysteries we uncover as we travel 13.8 billion years across space and time in search of our ultimate quest: the origin of the universe at the beginning of time itself.'

Danny peered at Bert through the opening in his helmet – the old man's eyes shining and his jaw suspended in a ridiculous grin. 'What about the ranch?'

'We are heading for the Missing Horse, but getting there is half the fun.'

'So what are we doing here?'

'We are going on safari,' said Bert, opening his door and climbing out of the Beetle.

'Oh Ma-an,' groaned Danny, and his head dropped to the dashboard in a puff of blue chalk.

3

Danny yanked out his mobile for another pointless search – moving it around his body and following it with his head, like a cat before it leaps at a fly. He shoved it back into his jeans and looked towards the road over the bush wall of the clearing.

'I have scouted a suitable area,' said Bert, appearing from behind the Beetle with a shiny, metal briefcase. 'Ah, we are the same height, five-foot-five.'

'I'm five-four.'

'It appears I have shrunk even more.' Bert placed the briefcase on the Beetle's bonnet and pulled two bottles of water from his shorts.

'Bert, how long are we going to be here?'

'Seventeen years,' said Bert, passing one of the bottles to Danny. 'Here's to Bert *and Danny's* Universal Tour and Expedition. It is now 1633 and our new understanding of relative motion and inertia is about to lead us to one of the most momentous discoveries in the history of science.'

'Why me?' muttered Danny, dutifully tapping his plastic bottle into Bert's.

'Do you not know?'

Danny tested with, 'Don't I know what?'

'Why I chose you to join me on my expedition?'

Fortunately, Bert thought he was asking a question. Unfortunately, Danny did know the answer. He nodded and leant back against the Beetle. It was his mum's fault. The first he knew about it was when she gave him the letter a month after choosing Bert's camp.

'Eh?' he said, sitting at the kitchen table as she cleared it of dishes and Alice filled it with chunky, felt-tip pens. '*Having considered your submission material, I am pleased to inform you that your application has been*

32

successful and Danny has been chosen …' He let the letter fall to the table. 'What submission material?'

'I sent them your Year Five science project,' said his mum, wiping her hands down her skirt.

'Dang it, mum!' he moaned, his arms falling down the sides of his chair. 'What did you do that for?'

He'd chosen 'THE SOLER SISTEM' but not wanting anyone to see it before it was finished and he'd shown it to his dad, he spent weeks in his bedroom pasting vivid drawings of the sun and her family of planets onto the large blue pages of his project book.

His mum dropped a pile of plates into the washing bowl from too high and adopted her don't-start-again face covered in white blobs. 'You should be pleased; you chose a very exclusive camp. And what are you worried about? Alice, don't draw on there darling. Your teachers, everyone agreed how good it was.' She puffed the foam from her cheek and turned back to the sink. 'How it showed your deep interest and enthusiasm.'

Everyone had also agreed it showed he didn't have a clue how to spell – and the title on the cover in big yellow letters, with flames licking off their sides, was just the start.

'Anyway, it worked, didn't it?' said his mum, her elbows working hard at her sides. 'You got the place you wanted. A deal's a deal. You agreed it's better than having to go to a camp back here when Tom's free.'

Tom was his best friend, but his addiction was burgers and Krispy Kreme doughnuts and he was being forced to suffer three weeks at fat camp.

'How many contitions are there?' said Alice, leaning closely over her colouring book, her tongue poking from the side of her mouth.

'It's con-D-itions and there's four,' he said. 'Two each.'

'That's fair then,' she said, with a single nod of her head.

'It's not a matter of fair,' said his mum over her shoulder. 'It's a matter of failing your physics exams for the second year in a row because you waste all your time shut in your room on that flipping computer.'

'How would you know what I do? You're never here. Or too busy.'

'Well you haven't spent it doing your schoolwork. And you don't lift a finger to help.' More foam blobs started flying as the dishes took the brunt. 'You don't go to basketball anymore, either at the club or the park.'

'I'm too short.'

'And you and Tom haven't been karting for months. It'll be Christine next. Then what?'

Christine ran the local Drama Club and was all waving arms and disorganized enthusiasm. Using words like penchant – and having one for improv and horror stories – she would shout out prompts at their impromptu performances; her hand rising into view above the stage, like *Thing*, to point frantic stage directions to the muddled cast. They were going to try Method acting, and there was no way he was giving up that as he planned to sleep on his bedroom floor and eat only baked beans for a week in preparation for the trip to Arizona.

His mum didn't know, but Tom could no longer fit in the junior karts, and you had to be sixteen to join *G-K Formula One*'s adult series.

Her elbows came to a stop and she took a deep breath and turned away from the sink. 'Danny, I know you don't want to go, but I think this will be really good for you. It's only a week and that will still give you two with us at Uncle Gil's.'

'I can't wait to see great Uncle Gil,' said Alice.

Great Uncle Gil was his dad's father's younger brother. He moved to America as a young man, made a fortune in securities and outlived them both. Alice had never met him but figured he must be a lot of fun to have it noted in his name. The family had visited him once when he was about four, but he only remembered him from the funeral – a bald head in a long black coat and an arm wrapped tight around his shoulders.

'You remember how much Dad loved your project?' said his mum, turning back to the bowl and pummelling the frying pan. 'What would he say?'

She only mentioned him when she was trying to get him to do something.

'I know one thing. I bet he would have chosen Bert's camp.'

'Daddy would say, you have to go,' said Alice, wagging a stubby finger at him. 'It's for your own good.'

'Yes, darling,' said his mum, using the back of her hand to push a strand of grey hair back to the messy bundle on her head, only for it to immediately fall again. 'Daddy would say that.'

Danny didn't want to, but he did remember.

He remembered his dad's praise, without a mention of spelling, and how they studied the clockwork model of the solar system on the table in the corner of his study, reciting 'My Very Excellent Mother Just Sent Us Nine Pizzas' as they named the planets in order from the sun. He remembered listening to the adventures of *The Durango Kid* from his Fifties Western comic collection, over and over again. And how, at Alice's age, he'd asked why the '*The*' was always said twice. He remembered creeping through the house, fingers twitching above his replica Smith & Wesson, ready to draw and shoot him wherever he lay in ambush, and performing staggering, chest-clenching throes after years of refusing to die. When all his dad could do was lie down, he remembered the comics had come back out and he'd read them back to him at his bedside, and his favourite novels like *Cimarron* and *Shane* ... His mum was right. His dad would have chosen Bert's Camp.

Danny jumped to attention. 'What?'

'Gravity!' said Bert, his arms wide and sunlight shining through his thin white hair, like a halo.

'What about it?'

'That is the momentous discovery.'

'Oh. Is that it?'

'You were expecting something more exciting?'

'Yes.'

'In cosmology, there is nothing more exciting than gravity. It is the weakest of the known forces but operates on a universal scale; the quest to understand it is the Holy Grail of science; and in 1633 Galileo quite literally got the ball rolling. At the time it was believed that a body's weight makes it fall because heavier bodies fall faster than light ones, but Galileo discovered something strange – something that is still very strange to most people. If you remove the effect of friction with the atmosphere and drop a hammer and a feather from exactly the same height at exactly the same time, they will fall side by side and hit the ground at exactly the same instant.'

Danny was in there with most people, but he took a large gulp of water to force his mouth to stay shut.

'And it makes no difference if one of the bodies is moving forward when it starts to fall. If I fire a bullet from a gun, level to the ground, and drop a penny from the same height as the barrel of the gun and at the exact moment the bullet leaves the end of the barrel, they will hit the ground simultaneously as well.'

Danny's head and shoulders jerked as he sprayed a fountain over the sand.

'Alright, I will show you.'

Bert tumbled the lock on the briefcase and opened it to reveal a replica of the Beetle's oval computer screen in the upright lid, and a keypad and Motorola 'brick' mobile from the 1980's nestled in the base.

'What are you doing with one of these?' said Danny, plucking the phone from its bed and inspecting the telescopic aerial zip-tied to its side as a replacement for the stubby original.

'My super-mobile,' grinned Bert. 'I needed the room to make a few improvements. It is really an Internet phone as it operates through my briefcase computer, but unlike most phones out here, it works.'

'Can I borrow it?'

Danny's focus switched as a whir of motors started deep within the briefcase. The computer beeped, '**searching...**' flashed onto the screen, and a metal bulb emerged from the top-right corner of the lid, rose eight inches on its aerial stalk and flowered into a small satellite dish.

'Wow!' said Danny, as the dish started to spin and tilt in precise movements. It finally locked into place, with a beep confirming a satellite link had been established, and a black-and-white, grainy astronaut bobbed onto the screen, moving in that otherworldly way that only comes from moving on another world.

'This is astronaut David R. Scott on the Apollo 15 mission to the Moon in 1971,' said Bert, leaving the soundtrack to take over.

'*There's something I think you'll find rather interesting,*' crackled Scott to Mission Control, '*which will only take a minute.*' He's standing in front

36

of the Falcon Command Module, the boulder-strewn surface of the Moon mirrored in his visor.

'Joe, I hope you have a good picture there. I've got –'

'Beautiful picture there, Dave.'

'Well, in my left hand,' continued Scott, *'I have a feather; in my right hand, a hammer. And I guess one of the reasons we got here today was because of a gentleman named Galileo, a long time ago, who made a rather significant discovery about falling objects in gravity fields. And we thought where would be a better place to confirm his findings than on the Moon, and so we thought we'd try it here for you.'*

The camera zoomed-in on Scott's heavily gloved hands, the hammer and feather pinched between forefingers and thumbs. *'The feather happens to be, appropriately, a falcon feather for our Falcon. And I'll drop the two of them here and hopefully they'll hit the ground at the same time.'*

The camera pulled back; Scott let them go; and falling together, they hit the Moon's dusty surface at exactly the same time.

'How about that!' cheered Scott.

'How about that!' came back Houston, applause ringing in the background as the picture froze.

'How about that,' Danny smiled, as though he'd watched a magic trick. 'How did Galileo discover it?'

'He rolled equally-sized but differently-weighted stone balls down a long, shallow slope so that the friction with the air and ground was the same for each and their reduced speeds allowed for accurate measurements.'

'They all rolled down together and hit the end at the same time?'

'That's right. But to measure this special motion, Galileo had to revolutionise seventeenth-century mathematics by introducing the concept of *time*. He realised that the speed of a moving body could be found by dividing the distance travelled by the time taken to travel that distance. So if we cover 100 miles in two hours, our speed for the journey was fifty miles per hour.'

'That's the average speed.'

'And that was the problem. As Galileo's balls rolled down the slope their speed was always increasing, which means the distance they covered in a certain time was itself changing over time. But how could Galileo measure the rate of change when the most accurate clock at his disposal was a sundial? His solution was to use a small, brass ball and weigh the passage of time using water. Opening a tap on letting go at the top of the slope, and closing it when the ball reached various distances along its length, he was left with an amount of water equal to the time it had taken the ball to complete each run. By weighing the water and using that as measure of time, he discovered that whatever distance the ball rolled after one second, it rolled four times as far after two seconds, nine times as far after three seconds, 16 times as far after four seconds and so on. In other words, falling bodies do not speed up randomly; they *accelerate at a constant rate*. In the same way that he had divided distance by time to come up with the formula for speed, Galileo divided the increase in velocity by the time over which the velocity is increasing to come up with the formula for acceleration. This showed that, in the absence of friction, every falling body continuously accelerates at the constant rate we call *g*, which is 10 metres or 32 feet per second multiplied by itself, or squared, for every second it is falling.'

Danny hadn't liked the sound of any of that. 'What about the bullet and penny?'

'Galileo realised that a body's velocity can have no effect on the constant rate at which it falls. Relative to you the penny is standing still in your fingers and starts to accelerate down at *g* as soon as it is released. But as it leaves the end of the barrel, the bullet is also moving at a constant velocity in its preferred state when it starts to accelerate down at *g*.'

'Oh, it can be said to be standing still as well.'

'Exactly. You think the bullet falls more slowly than the penny because it goes shooting off before either of them have had time to fall very far, but they were released at the same time from the same height and will accelerate down at the same constant rate, regardless of their relative motion.'

'It still sounds weird.'

'Get this. If you jump-up on the spot, as high as you can, and reach a height of three feet, then take a run-up and long-jump, as far as you can, but reaching the exact same height in the midpoint of your jump, in both cases you will remain in the air for exactly the same length of time.'

Danny laughed.

'Gravity acts at the same constant rate for all bodies, so the only factor determining the amount of time in the air is the height above the ground. How the body is moving at the time is irrelevant. With this new understanding, Galileo discovered that if we combine the two paths of any jump, throw or launch on a graph, plotting the body's velocity at the instant it is released against g, the constant rate of fall due to gravity, every body follows a curved flight path called a parabola.'

'I've heard of that. What's a parabola?'

'A parabola is the curve produced by slicing into a cone so that the knife dissects the base. Just as there are any number of different ways to slice a cone, so there can be any number of different-shaped parabolic curves, but Galileo was able to predict the parabola a body would follow based only on its initial trajectory. He advised the Italian Army how to improve the accuracy of their cannon fire, demonstrating that to achieve the greatest distance they should fire at an angle of 45 degrees to the ground.'

'That's what we're told to do with the javelin at school.'

'And it is all because everything falls at the same constant rate.'

'Yeah, but *why* does everything fall at the same rate?'

'Why does anything fall at all? Galileo didn't know. He was only trying to measure how bodies move and did not even think of gravity as a force. No, the full significance of a body's preferred state of motion was not discovered until a seventeenth-century countryman of yours sat under a tree and an apple fell on his head.'

'Sir Isaac Newton.'

'Probably the greatest scientific mind the world has ever seen, he worked out three simple laws of motion that made sense of all the movement in the heavens, and all the wild, apparently random movement here on Earth. Armed with these, he discovered gravity and the universal law of Nature and mathematics that it follows. So, come on, let's put this away and find out how he did it.'

Bert held out his hand, and realising he still had hold of the Motorola, Danny pulled it out of his reach. 'Bert, can I make a quick call?'

'Yes, of course. The extended aerial gives the handset a range of around 100 feet.'

Danny pulled out the three-foot aerial, and punching in the number his mum had given him to use in emergencies, sauntered out of the clearing. Surrounded by dense scrub he dropped into a squat as the ringing stopped, but it was just an answering machine kicking in.

'Mum!' he whispered. 'You got to get me out of here. This camp's not what we thought it was. It's just me, some old guy pretending to be Einstein and a pompous Frenchman, but I haven't met him yet. I'm pretty sure he's not following proper Health and Safety guidelines.' The other camp brochures had whole sections on that. 'He's already tried to do me in twice!'

Danny popped up and peered over the bushes at Bert half-buried inside the Beetle. 'With no other students to bounce off,' he said, ducking back down, 'I'm not going to learn anything that'll help my grades. We're not even following the syllabus. Hang on.' Bert was heading across the clearing in his direction. He spoke quickly. 'Reply to this number; my mobile isn't working here. I've got to go. He's coming after me again.'

Flicking-out his right foot and bending at the knees, Danny started thrusting and parrying the aerial; grunting and scuffing his feet as he leapt around, like d'Artagnan fighting the Cardinal's guards. Turning the aerial on himself, he dropped to his knees; gasped as he drove the telescopic steel into his stomach; held the phone in the silent air for three seconds then pressed the cancel key. Jumping to his feet, he headed for the Beetle and passed the telephone back to Bert as they met at the edge of the clearing.

'Is everything alright?' said Bert, a fringe of white hair sprouting from beneath a khaki pith helmet.

'Yeah. Fine thanks,' panted Danny. 'Just leaving a message for my mum ... tell her I got here safe and sound.'

4

It hadn't registered with Danny when Bert first mentioned going on safari, but as the old man grinned beneath his pith helmet, and the sun beat down from almost overhead, a second helmet appeared from behind his back and landed on Danny's head.

'What's this for?' said Danny, eyeing Bert suspiciously.

'We are going to follow in Newton's steps and track down an elusive beast. It comes in many guises, hiding its true nature, so finding it will be tricky enough. Once we find it we have to capture it.'

'Can't we just shoot it?'

'Ha! No, the beast we are looking for is a *force*. We have talked about gravity, friction, engines and we could name many more, but what connects them? What is the real nature of a force? ... I think it is this way.' Bert fished his pipe from the breast pocket of his khaki shirt and marched straight past the Beetle, his back bent forward as if by purpose more than age.

'Where are we going?'

'To see what got Newton started,' called Bert, weaving between the bushes at the edge of the clearing. 'Come on.'

Danny trudged after him, thinking about his mum and getting further and further behind. By the time he'd reached the conclusion that she was never going to buy it, he'd caught up.

'It is now 1650,' said Bert, startling Danny to a stop in front of an expanse of low sagebrush and open sand. 'Newton is about to conquer the cosmos and it all starts with the discovery of a force.' He pulled a red, tennis-sized ball wrapped in string from the depths of one of his pockets and offered it to Danny.

'What's any of this got to do with us being out here?' said Danny. One hundred yards away the Beetle's loaded roof rack and rocket-thruster dustbins hovered above the clearing, like the Starship Enterprise.

'There are fewer bushes over here.' Bert poked the end of his pipe into Danny's ribs. 'And you will enjoy this. We are about to remove another blindfold.'

Danny snatched up the wooden ball and let it roll down the three-foot length of string until it dangled from his hand, like a badly thrown yoyo.

'This red ball is *Io*, Jupiter's fifth moon. Start twirling her around your head.'

'I know this,' said Danny, winding Io into motion.

'Try and keep her moving at a constant speed … Say that is 20 mph. What do you feel?'

'It's that what-do-you-call-it force pulling on me,' shouted Danny, his hips rocking back and forth to balance against it.

'It feels like there is a continuous force trying to pull Io out of your hand. What happens when you increase the speed?'

'The force gets stronger,' said Danny, powering Io to twice the speed in a blurred orbit around his fist.

'Okay, so if you slow her back to 20 mph … Now, what will happen if you let go?'

'She'll fly off.'

'In which direction?'

'It depends when I let go of the string.'

Danny let go, with a flourish of the wrist, focusing on a soaring flight down the imagined fairway in front of him as Io flew off to the side, narrowly missing Bert.

'That's right,' said Bert, climbing back to his feet. 'The ball appears to fly off in a straight line from wherever it is in its orbit at the instant you let go. In 1650 a Dutchman called Christiaan Huygens dubbed this effect *centrifugal force* from the Latin for fleeing the centre.'

'That's it. Centrifugal force. I've been on one.' It was at a steam-driven fairground the previous summer. 'You know, you stand against the wall in a round room then it starts spinning and the floor sinks away but you stay glued to the wall in mid air.'

'It is called a centrifuge,' said Bert, pulling a second, larger, green ball from his other pocket.

'I think it was Mr Boggle's Sticky Wall,' Danny mumbled, noticing that the ball's removal had done nothing to alter the shape of Bert's shorts.

Bert passed over the green ball. 'Today you are our centrifuge. This is *Ganymede*, Jupiter's seventh moon. She weighs twice as much as Io but try and keep her moving at the same 20 mph ... What does that feel like?'

'It feels a lot stronger than Io,' shouted Danny. 'Or about the same as when she was going twice as fast.'

'What happens when you let go?'

Bert tucked in behind Danny as he let Ganymede fly, and together they watched her soar down the middle of the fairway. 'That was a good one.'

Danny had expected it to be much better.

'So, it appears that increasing the *speed* of rotation or the *weight* of the body increases the size or magnitude of the centrifugal force, which means –'

'The centrifugal force is the body's weight multiplied by its speed,' pounced Danny.

'The scientific term for the matter in a body is *mass* not weight, although historically you are correct because it is now 1650 and Newton did not introduce the term until 1687. But mass and weight have the same value here on Earth so we will use the technically correct mass.'

'What is the difference between weight and mass?'

'We will find out in 1687. But your use of the term speed is also incorrect because the ball can fly off in any direction, depending on when you let go of the string.'

'Then it must be its weight – I mean mass times its velocity.'

Bert grabbed Danny by the shoulders, his face a mass of grinning crinkles. 'Have we done it?' He proceeded to give Danny little shakes for each part he wished emphasised, which turned out to be all of it. 'Have we tracked down and captured that centrifugal force is the force of a body when it is moving in a circle and at any instant the size of the force is the body's mass multiplied by its velocity at that instant?'

'Yes,' Danny nodded involuntarily, his pith helmet a beat behind. He'd thought it couldn't be that hard.

'No.' Bert dropped his arms to his sides and set off after Io. 'You go and get Ganymede and I will meet you back at the Beetle.'

'What do you mean?' said Danny, double-timing to catch up.

'I mean, whilst I retrieve Io, you go –'

'Noo! Why haven't we tracked down a force?'

'Ah, well, the thing is, centrifugal force is not a real force at all.'

*

'What do you mean centrifugal force is not a real force?' said Danny on arriving back at the Beetle.

Bert handed him the open briefcase. 'There is a message for you.'

Drawing the case close to his chest, Danny turned away and opened the SMS.

> Nice try and so soon.
> Can still get you on the Cool
> Science 2-week intensive
> course here in Phoenix. Or
> there is always Camp Boffin
> when we get back. Is that
> what you want? Have fun.
> Love mum xo

Chewing his bottom lip, he deleted the message, passed the briefcase back to Bert, and watched the satellite dish fold its filigree petals and retract into the lid.

His chances of successfully getting out of there were now seriously limited – successfully being his mum not finding out if he managed to bunk off this camp so that he wouldn't have to go to another. He needed to come up with a plan but figured his best bet was to get to the Missing Horse and the privacy of his own phone.

'Jump in,' said Bert, placing the briefcase behind his seat and climbing into the Beetle. 'Now if centrifugal force is not a force –'

'Yeah, why isn't it a force?' said Danny, hastily fastening his seat belt as the Beetle pulled out of the clearing and the bush walls of the track closed in on either side.

'It was called a force through a lack of understanding.'

'But I could feel it pulling on me and it made the balls fly off.'

'What did you notice about your two throws? They were rotating at the same speed when you released them, but though the heavier Ganymede produced twice the centrifugal effect on your body, she flew along a similar parabolic path and over the same distance of ground as the lighter Io.'

Bert was right, Ganymede hadn't flown as far as he'd expected. 'Why didn't she go further?'

'That's what people wondered in 1650, but they were confused by Galileo's idea of inertia. He realised that a body's preferred state of motion is to keep moving at a constant speed in the same direction, but he thought the same direction was a circle.'

'Eh?'

'It does sound crazy but Galileo reasoned that if you remove all friction, a freely moving body on the surface of the Earth would keep moving in its preferred state and eventually travel all the way around and complete a circle.'

'Oh yeah, it would wouldn't it.'

'No it would not, and that is what Newton realised. A body's preferred state of motion is how it will move when there are *no* outside forces acting. But when Io and Ganymede were orbiting your fist a force was acting: you and the string, forcing them to constantly change direction. It was only when this outside force stopped acting and you let go –'

'Oh-h, the balls moved into a straight line because they were going back to their preferred state.'

'Exactly. Centrifugal force is the effect a body experiences when it is forced to turn or move in a circle because at every instant it wants to return to an inertial frame moving in a straight line. Ha, ha!' laughed Bert, bouncing off his seat. 'That is Newton's first law of motion, which simply states that *every body in the universe will remain at rest or in uniform motion along a straight line unless acted on by an outside force.* Newton introduced the term *momentum* as a measure of this natural inertial motion, represented by the letter p, and that is what mass multiplied by velocity measures in the equation $p = mv$.'

'It was my momentum that kept me going into the windscreen.'

'In the same way, when you jump straight up in the aisle of a moving bus, the bus does not carry on beneath you because you are sharing its motion and your momentum keeps you moving forward at that same velocity for the entire time you are in the air. Like the ping-pong ball bouncing in my Beetle, to an observer on the bus you jump straight up and down to land on the same spot as you took off. And to an observer at the roadside?'

'I jump forward with the bus to land on that same spot as well.' Danny had tried and always wondered about that.

'And because momentum is mass multiplied by velocity and one is directly proportional to the other, the momentum of Io moving at 40 mph was the same as the twice-as-heavy Ganymede moving at 20 mph and produced the same size centrifugal effect on the force of you twirling them around your head. When you released them, each moving at 20 mph in a circle, they settled into an inertial frame moving at 20 mph in a straight line, falling at g from the moment you let go and –'

'Travelling the same distance before hitting the ground.'

'That is Newton's first law of motion.'

Bert was driving at a sensible pace as the Beetle approached the end of the track, but they emerged onto a narrow strip of sand at the edge of the tarmac to find a huge articulated lorry roaring towards them; its horn a rising scream to keep out of the way – which Bert did by immediately pitching them to a stop.

The lorry shot past, rocking the Beetle with the air it had pushed out of its way, and rumbled on down the road with its engine deepening, like distant thunder.

'What sort of speed's he doing?' said Danny, as the Beetle crossed the road and started accelerating the other way.

'**72 mph east**' popped onto the computer screen, jumped straight to '**73**' and started working its way up.

'How does your computer know that?'

'There is a police radar gun fitted to each side of the roof rack.'

Bert's grins were always at their best when he talked about his Beetle, and Danny smiled as the white hairs sprouting from his ears all stood to attention. 'How does it know when you're speaking to it?'

'My speech-recognition software identifies specific command

words but will only accept them from an authorised source.'

'So it won't listen to me?'

'It just did,' said Bert, popping his pipe into his mouth. 'It has been learning your speech patterns all morning.'

'Are you ever going to light that?'

'Unfortunately, no. At my age it is a luxury I can no longer afford. I tried lollipops but got fed up with them.' Bert opened the glove box to a multicoloured hill of interlocked balls. 'Besides, I enjoy playing with my pipe. It helps me to think straight.'

Danny tried again. 'How old are you?'

'Older than I should be for my age. Now ... Where was I?'

The road charged to the western horizon, as though governed by Newton's first law, and as the Beetle rolled along it with no apparent input from Bert, Danny sucked on a strawberry crème and let his eyes drift across the patchwork green plain to a range of red sandstone hills shrinking in the rear window. 'This is the same stretch of road!'

The computer beeped and '**30 mph west**' appeared on its screen with '**cc**' blinking underneath.

'Ah yes, Newton's first law of motion,' said Bert. 'If Newton is correct and every body possesses a natural straight-line inertia to keep moving at a constant velocity, what must be happening with our Moon or any orbiting body?'

'There must be an outside force keeping them in orbit,' patted out Danny, raising his eyes and rocking his head from side-to-side. 'Can't we go a bit faster?'

'There must be an invisible force, acting like a piece of string, continuously forcing the Moon to change direction.'

'Yes, gravity, but can we –'

'Remember, it is 1650 and Galileo's circular-inertia already explains an orbit so no one is even looking for a force let alone thinking it could be gravity. No one except Newton.' Bert gave Danny a nudge, as though he'd recommended a film and they were getting to the good bit. 'When you hold a penny in your fingers it is at rest in its preferred state. When you let go, some kind of invisible force makes it move relative to you and keeps it accelerating towards the ground.'

'Gravity,' Danny droned again, dropping his head to his chest.

'So Newton had discovered two hidden forces as a consequence of his first law. His genius was to see that each could be a different aspect of the same force. It did not land on his head, but one day, watching an apple fall from a tree, it occurred to him that if this invisible force could reach to the top of a tree, could it reach even higher to pull the Moon in its orbit around the Earth. To work it out he would have to measure a force, and that is what we are going to do. So, come on, what is a force?'

Never mind what it was, Danny just wished more of it were coming out of the engine. 'I don't know,' he said, raising his head and waving his arms, like a puppet in the control of unseen hands. 'Friction's a force, gravity's a force, your Beetle driving along now's a force. Who cares? I want to ride a horse.' He flopped down again, as though someone had cut his strings.

'No it's not.'

'What?'

'My Beetle driving along now is not a force.'

'Yeah right,' scoffed Danny. He'd seen enough safety films of people being hit by them to know that a moving car is definitely a force.

'According to the mathematics, it cannot be.'

'Oh Ma-an, not maths as well.'

'Ha! No, you see a single body moving at a constant velocity is in a preferred inertial frame, which means it is equivalent to standing still with zero velocity. Anything multiplied by zero equals zero, so it cannot be a force.'

Danny had every intention of remaining silent but Bert's explanation forced out a guffaw and, 'What sort of weird sense is that! We've got weight, I mean mass, and we're moving so we must have momentum. And I know Ganymede didn't fly further but I still can't see why that isn't a force.'

'Alright,' said Bert, pressing a sequence of keys. 'If my Beetle is a force as we drive down this road, we should be able to measure that force. What do you and your rucksack weigh?'

'**62.93 kg**' beeped onto the computer screen.

'How does it know that?'

'Sensors on the shock absorbers register the weight of the car body and since your arrival it has increased by almost 63 kg.'

'My rucksack's 13 kg. They weighed it at Heathrow.'

'Together with your 50 kg that makes my Beetle's overall mass, or weight, 2003 kg, which we can round down to two tonnes. Now, suppose we are in outer space.'

As the black screens rose silently up the windows, Danny's hands accelerated faster than gravity down his sides. 'WHOA!' he cried, pushing back into his seat and clutching the bottom of its frame.

'Here we are back in Spaceship Beetle,' came Bert's voice from the darkness, 'all alone, out deep in empty space. We know our mass is two tonnes, what is our velocity?'

'It's 30 mph, Bert!'

'We can come up with any velocity we like from some other point of view. But we are all on our own and from our point of view ...'

The computer beeped '**0 mph**' onto its screen and a green glow onto the edges of everything else.

'And there's traffic about!' cried Danny.

'If our velocity is zero,' continued Bert, 'then we have zero momentum and our inertia cannot be a force.'

Danny relinquished his handholds for a frantic attack of cinching; pulling the seat belt tight across his stomach and up across his chest only to discover a large loop draping over his fist and refusing to wind back into its spool. 'Oh-h. What if we crash into something?'

'Exactly,' said Bert, punching Danny full on the arm. 'A body's momentum on its own is not a force just as a fist travelling through the air is not a punch.'

Bolting upright, his hands back on the seat frame and his face contorted in anticipation of the rest of the impact, three things occurred to Danny. First, his seat belt had snapped firmly across his chest. Second, there hadn't actually been an impact. And third, Bert had a point. 'Oh yeah,' he said, rubbing his shoulder. 'You only feel the force when you hit something.'

'And there has to be a something else to hit. We have picked up the scent. Like all measures of motion, momentum is a relative measurement and we need the momenta of at least two bodies interacting in a collision to produce a force.'

'Yeah, and if we do hit something we're going to deliver a big one.'

'No we won't.'

'Be-beep, be-beep' … 'Be-beep, be-beep' … 'Be-beep, be-beep'

Danny froze. 'What's that noise?'

'What noise?'

'It's like a low beeping … getting louder.'

'Oh, the collision alarm.'

'The collision alarm! Dang it, Bert, clear the wind –' Danny stopped suddenly and his hands relaxed their grip. 'We've rolled to a stop again, haven't we? Yeah we must've done or you'd be worried we'll hit something.'

'How can we hit anything when we are standing still?'

'Then why's the alarm going off?'

'That alarm is not warning that we are going to crash into something.'

'Oh.'

'It is warning that something is going to crash into us.'

'What difference does it make!' exploded Danny. 'Clear the windows.'

'Hmm, that's true. Either way we are going to receive a force. That must be it. A moving body in a collision does not deliver a force, it only receives one.'

'What's that mean?' said Danny, pointing to the computer.

On screen, '**30 mph east**' flashed on-and-off in large red letters and in time with the be-beep of the alarm.

'That means something is approaching, head-on, at 30 mph.'

'We must have crossed over the road,' said Danny, back on full alert.

'Ha! Do not worry,' laughed Bert. 'Look.'

A blob of light had appeared in the middle of the black windscreen, heading straight for them and growing fast.

'It's Spaceship Beetle II again, but this time on a collision course. We can now measure that she is moving at a velocity of 30 mph east.'

Danny shook his head and smiled. 'Or we are moving at 30 mph west.' He watched Spaceship Beetle II close to within a few hundred yards, the bug-shape clearly defined under the load on its roof.

'And because we are travelling along the same straight path,' said Bert, 'in a minute we are going to collide.'

'Egzackerly. We must hit each other and –'

Danny pitched violently forward, his head rocking off his chest and only his seatbelt stopping a reunion with the windscreen – a fate neither his pith helmet nor lollipop avoided. He looked at Bert with the wide eyes and mouth of the totally shocked, and Bert looked back with the raised brows and half-smile of the mildly surprised. 'What was that?'

Danny jumped at his own voice. They now sat in a silence more sinister than any alarm; the computer screen was an oval mass of wavy, grey lines; and the windscreen was completely black again ... Spaceship Beetle II was nowhere to be seen.

'Bucking broncos, Bert! We've hit something!'

5

The Beetle's roof light blinked on, bouncing pale-yellow beams off Bert's lenses as he exhaled and gave them a polish with his shirt. 'You mean something hit us.'

'Seriously, Bert. We've hit something.'

'How can you tell?'

'Because, like my hat, I went to hit the windscreen again, that's how. We must have been moving, hit something and my momentum kept me going forward when the car slowed down.'

'Yes, very good, but what would your body do if we suddenly accelerated backwards?'

It didn't take Danny long. 'I'd do the same thing.'

'As we saw when we accelerated to a stop on the road, any movement in one direction can be seen as the same movement in the opposite direction from another point of view.'

'We were slowing down from the Earth's point of view and speeding up from the North Pole's.'

'So rather than argue that we were moving due west, hit something and slowed down, we can as easily argue that we were standing still and something hit us head on, which made us accelerate due east. And your pith helmet and head did not keep moving forward as the Beetle decelerated; they were left behind as your seat, with your body strapped to it, accelerated backwards. A split-second later your body forced your head to do the same, and a split-second after that the windscreen forced your helmet.'

'Okay, something crashed into us. Just clear the windows.'

'Ha, ha!' laughed Bert, pointing at Danny and dropping his head to his chest, which was as much collapsing as he could manage in his seat belt. 'Do not worry, we did not really hit anything.' He paused for an attempt at some rocking. 'No, that was my Beetle automatically braking to simulate being hit by Spaceship Beetle II.'

'She was still miles away.'

Bert dropped the grin and went for the keypad. 'Yes, it was a little out of sync. There must be a glitch in the program.'

'That didn't feel like braking. It happened so fast.'

'It worked well, didn't it? Better than in any of my tests.' Bert studied the lines of computer code scrolling up the screen. 'This is the MFDP, my Beetle's Master Fault-Diagnostic Program. The problem has something to do with the 30 mph collision warning … it appears to be stuck.'

'Hold the horses. If you braked to simulate us being hit by Spaceship Beetle II, then we were moving.'

'Yes, but not for long. As soon as we entered spaceship mode, the CSP, my Collision Simulation Program, started to gently slow us down and the full application of the brakes bought us to a stop.'

'Relative to the road?' Danny wanted it posted on Twitter. 'We are not moving on the road?'

'No,' said Bert, dismissing the idea with a wave of his hand. 'We can forget about Spaceship Earth because, out here in space, there are only our two Beetles.'

'No there aren't.'

'I know,' said Bert, his fingers bouncing off the keys. 'There is a problem with the software but I cannot locate it. Spaceship Beetle II should be large as life in the windscreen.' He glanced up to check if it was, and then tried a thump. 'Oh well, we will have to imagine it is there. Now, what can we measure? From our point of view we were standing still, Spaceship Beetle II crashed into us at a velocity of 30 mph east and we received a force.'

'Then Spaceship Beetle II can't deliver a force either.'

'That's right. From their point of view their momentum was zero and cannot be a force. The maths seems to be telling us that a collision cannot take place.'

'Duh!'

'So what is the logic of the maths trying to tell us? Neither Beetle can use its individual momentum to come up with the force, but what happens when two momenta meet and resist each other, like in our collision?'

Bert pressed a key on the console and settled back in his seat, until '**0 mph**' appeared on the computer screen and he sat forward again. 'Drat,' he said, pushing the key repeatedly with the same result. 'The speedometer should read that we are now travelling backwards, heading east, but nothing is coming back from the MFDP.'

'What about clearing the windows?'

'Good idea,' said Bert, selecting a new key. 'The program just needs a little time to find the problem.'

Danny watched Bert's bony finger progress from a simple push to a frenzied stabbing, like the *Psycho* in the shower scene.

The Beetle's windows remained impenetrably black.

He tried his door.

'They are automatically locked in spaceship mode,' said Bert, furiously typing.

The computer beeped and '**20 mph west**' appeared on its screen.

'What's that mean?' said Danny.

Bert's face glowed sickly-green in the light from the screen as 20 mph west kept reappearing across it. 'The crash program should have slowed us down to 10 mph west and then braked us to a halt. Somehow the computer doubled the speed of the collision.'

'Can it do that?'

'No.'

'So what's happened? … Bert, we've got to clear the windows.'

'I know, I know,' said Bert, unclipping his seat belt. 'I am going to reboot the entire system. Let me know if anything happens.' Turning his back to Danny and placing one knee on the floor and the other on his seat, he slid under the steering wheel to access the Beetle's fuse-box. 'What about now?' he grunted.

'No.'

'And now?'

Feeling a tapping on his bottom, Bert twisted round to find the Beetle flooded with sunlight and Danny's finger still tapping in mid-air. He elbowed his way between the steering wheel and the door, and with a 'Yaa-ah!' sprang the rest of the way into his seat.

The Beetle stood in the middle of its lane, with no one in sight and everything perfectly still, silent and just as it should be, except for one thing. Unfortunately, it was a large, maroon, crumpled thing that began before the front of the Beetle ended.

'What's that doing there?' cried Bert.

'You smashed into it, that's what.' Danny stared at the cracked windscreen perched on top of the Beetle's crash bars, and the yellow smiley hanging motionless from its rear-view mirror, telling him to *Have a Nice Day*.

'But that's Morris' Citroën 2CV.'

'Your housekeeper?'

'Yes. I told you it was stolen last night. It must have hit us at 30 mph and caused the software to crash.'

'We were the one's moving.'

'For the computer-simulated crash, yes, but if the radar identified a possible real collision, the emergency override would bring us out of spaceship mode and to a stop.'

'That must be the problem then.'

'No-o,' said Bert, lowering his wiry eyebrows and shaking his head. 'That is a fail-safe. We cannot have hit Morris' car.'

'Then where's the driver?'

'Maybe he ran off. It is a stolen vehicle after all.'

'Ran off!' said Danny, holding his hands up, like scales, and looking from left to right.

He climbed out of the Beetle, scanning the open plain from behind the shield of the passenger door. The dense brush had thinned into a threadbare carpet of desert scrub that offered little to conceal an angry thief, but he reversed around the Beetle's boot, like the Rango Kid backing out of a saloon, before turning his attention to the wreck.

The Citroën's sloping boot and rear wheel arches gleamed, as though they'd been polished that very morning. But its front axle and engine had been rammed beneath a buckled cabin, leaving the remains of its crumpled bonnet wedged into the Beetle's crash bars and only its back wheels touching the ground. The only damage to the Beetle was the loss of dried mud from around the Citroën's steel claws, revealing bright-yellow paintwork beneath.

'Morris was ever so fond of that car,' said Bert, joining Danny at the side of the Citroën and frisking himself for his pipe. 'He had it specially imported.'

Danny peered at the twisted steering column through the shattered passenger window. 'Hey, the keys are in the ignition.'

'Ah! There you go then. I told you the Citroën crashed into us.'

'That doesn't prove anything?'

'Well, no.' Bert shifted his feet in the sand. 'But our maximum speed can only have been 10 mph.'

'The collision was too violent for that. We were doing thirty.'

'Then it must have been a 30 mph impact by the Citroën. That is what the computer was warning us about, not the simulated crash.' Bert's face managed to sag a little further and he turned and trudged back to the Beetle. 'Either way, I had better phone Morris.'

As the old man pulled the briefcase onto his lap, Danny decided to investigate the scene of the accident. He inspected the road behind each car, but there were no skid marks and, weirder still, though the Citroën's windows and frog-eyed headlamps were clearly smashed, there was no broken glass or debris anywhere.

'I have left a message,' said Bert, climbing out of the Beetle. He reached through the Citroën's passenger window and removed a white paperback book from the glove box. 'We are going to have to wait for Morris to arrive, so if we want to find out what happened we will have to look at the maths and see what we can measure.'

'What do you mean, look at the maths?'

'Do not panic, you have been able to do this maths since you were a child. Let us roll up our sleeves, gird our pork and see where it leads us.' He stopped. 'While we are at it, we should have lunch. Grab the chairs off the roof rack and I will get the sandwiches.'

'Oh,' said Danny. 'Alright then.'

*

Lunch was served from a Tupperware box sat on top of a large, blue cool-box, with a collapsible metal chair either side, and facing a bug-like Citroën climbing up the front of a Beetle, as though ineptly trying to mate. Looking east, back the way they had come, the range of hills had finally disappeared, and looking west, the road continued its relentless run for a few miles then disappeared at the base of a north/south-running ridge.

'Okay,' said Bert, after washing down the remains of his first sandwich. 'We know a body's momentum cannot deliver a force on its own, but can we use the momenta of two bodies in a collision to measure the size of the force each received? What was the momentum of my Beetle and Morris' Citroën before the collision?'

'That's the point,' said Danny, around the last of his third sausage sandwich. 'We don't know how either of us were moving.'

'But if we say the collision speed was 30 mph, which I think is a safe assumption given the damage to Morris' car, that is a constant that cannot change regardless of our relative motion.'

'Oh yeah, that's the same from every point of view.'

'And we know our respective masses.' Bert waved the white owner's manual from the Citroën's glove box and started flicking through its pages. 'My Beetle weighs two tonnes and the Citroën weighs ... good. We can round it up to one tonne.'

'Let's see,' said Danny, leaning over the page of stats to read 'Unladen weight, 998 kg'.

'First we will look from our point of view in my Beetle. We can only measure that we were standing still and the one-tonne Citroën crashed into us with a relative velocity of 30 mph east and a momentum of ...'

'One x 30 –' Danny stopped. 'Thirty what, east?'

'Momentum is measured in *kilogram metres-per-second*, which means how many kilograms are travelling how many metres in each second that passes from a specified point of view. But the units are not important for our purposes, so to keep the maths simple let's say the Citroën has 30 units of momentum or UM. What was our momentum in my Beetle?'

'*If* we were standing still then it's zero.'

'Which means we cannot have offered any resistance or had any effect on the Citroën's momentum and there must be 30 UM heading east after the collision as well. Again, the maths is saying it is as though it never happened.'

Danny emptied his bottle of cola and pointed it at the two cars, separated only by colour. 'We know it happened and we know it had an effect.'

'But what do you notice? Before the collision the cars were

moving along the same straight line at different velocities. After the collision they have become *one body* travelling in the same preferred state along that line.'

'You mean they're standing still relative to each other and the ground?'

'They are only standing still relative to the ground because of friction with the Earth and its atmosphere. If they were out in space, the Citroën would have picked up the stationary Beetle and taken her with it, combining to form Spaceship Citroën/Beetle moving as one at a new constant velocity.'

'Oh, but now it weighs three tonnes.'

'That's right, and each tonne is sharing the total momentum of 30 UM heading east. So in our momentum equation, $p = mv$, if Spaceship Citroën/Beetle's mass is three tonnes and the answer is 30, then their combined new velocity must be …'

'Ten mph east.' Danny imagined the collision taking place on the road in front of him – the Citroën ploughing into the stationary Beetle at 30 mph from his right, immediately slowing by 20 mph and accelerating the Beetle to 10 mph, then continuing together at that speed down the road to his left.

'My Beetle's momentum was zero UM before the collision. What is it after the collision?'

'We make up two of the three tonnes moving at 10 mph east, so it must be 20 UM east.'

'So let's say she received 20 units of force or UF heading east. The standard unit of force is the *newton second*, which is the force needed to accelerate one kilogram of mass by one metre per second every second. But again, to keep the maths simple, we will say my Beetle received 20 UF, which accelerated her two tonnes from zero to 10 mph east. How has the Citroën's momentum changed after crashing into us?'

'Its one tonne is only a third of the total 30 UM heading east, which is 10 UM east. So the Citroën lost 20 units of force east.'

'And that force caused twice the change in velocity to the Citroën because it has only half the mass or weight of the Beetle.'

'The Beetle accelerated by 10 mph and the Citroën slowed down by 20,' nodded Danny. 'But you said a body only receives a force from a collision and yet the Citroën lost 20 UF.'

'Do not forget, all movement is relative. What would we have measured had we been sitting in the Citroën for the collision, ignoring the fact that we would have seriously increased its mass or weight?'

'Now we're getting to the truth.' Danny started checking the maths through in his head. 'If your Beetle crashed into the Citroën with a velocity of 30 mph west, the total momentum after the collision was 60 UM moving each of Spaceship Beetle/Citroën's three tonnes at a velocity of 20 mph west ... So the one-tonne stationary Citroën gained 20 UM east and received 20 UF, which made her accelerate backwards from zero to 20 mph west. And your two-tonne Beetle lost 20 UF west and decelerated by 10 mph to 20 mph west as well. It's the same size force for both of them but in opposite directions.'

'Exactly. Gaining 20 UF and accelerating in one direction is completely equivalent to losing 20 UF and decelerating in the opposite direction, but we measure both bodies to receive the same-size force regardless of which point of view we take. We have finally tracked it down. In a collision the bodies' momenta resist each other to produce one-size resultant force and that force acts on both bodies in opposite directions.'

Danny lifted the lid off the cool-box, pushing past potatoes, bagels and an oven-ready chicken to pull a second bottle of cola from the ice. 'Is it the same no matter how many bodies collide?'

'Provided they all hit at the same instant, yes. There will be one resultant force and each body will receive that force as though they had been standing still and resisted the combined momenta of all the others. Now the maths is starting to make sense. One body on its own cannot be a force because a force is the result of a collision, not the cause. And the size of the force is its effect on the body that receives it because that is all we can measure.'

'We've done it then,' said Danny, searching through the Tupperware box for something less savoury. 'The force is the change in momentum.'

'What is a change in momentum? What did my Beetle do if we say the Citroën hit us?'

'We accelerated backwards, I mean east.'

'What did the Citroën do if we hit it?'

'It accelerated west. You mean the force makes the body accelerate in the direction of the force?'

'That's right. We also know that the same size force had twice the effect on the Citroën, so what would happen if we each received 40 UF?'

'Your Beetle would accelerate by 20 mph and the Citroën by 40.'

'So the amount of acceleration caused by the force is directly proportional to the size of the force. If you double the force you will double the amount of acceleration; if you triple it, the acceleration will be three times as great and so on. And as Galileo discovered with gravity in his rolling ball experiments, if the original force acted on my Beetle continuously, we would keep accelerating with our velocity increasing by 10 mph for every 20 UF we receive.'

Water slopped over the lip of Bert's water bottle as he threw his arms wide. 'That is Newton's second law of motion, which simply states that *an unbalanced force applied to a body will cause it to accelerate in the direction of the force and the amount of acceleration will be proportional to the strength of the force.*'

'Is that it?'

'That's it. So the simplest way to measure the size of the force is to multiply the body's mass by the acceleration it experiences. From our point of view that was 2 x 10 mph for my Beetle and 1 x 20 mph for the Citroën, in both cases giving us 20 units of force. Ha, ha! We have tracked down a force with Newton's second law of motion and captured it with his force equation, $F = ma$.'

'Yeah, but which point of view is right?' Danny was about to remind Bert that they were dealing with real cars on a real road, not virtual spaceships in his computer, when he discovered, hiding at the bottom of the Tupperware box, peanut butter and jam sandwiched between what could almost be described as white bread.

'You mean what happened in our collision relative to Spaceship Earth?' said Bert. 'In the same way that our collision speed has to be 30 mph from every point of view, and we use the addition and subtraction of velocities to work it out, so the momenta measurements must always battle to produce the same force of 20 units because that event cannot change either. The only difference is momenta add and subtract the other way round, like forces.

'Equal and opposite forces cancel out,' said Danny. 'So you … subtract momenta moving in opposite directions and add momenta moving in the same direction.'

'Which means if two identical cars collide right in front of us, each moving at the same speed but in opposite directions, there would be an almighty crunch but from our point of view on Spaceship Earth their momenta cancel out and there would be no resultant force.'

'The cars receive a force.'

'Because from their point of view their momenta were not equal and opposite. Similarly, from the point of view of Spaceship North Pole the cars were moving in the same direction; you have to add their momenta; and there is a positive resultant force. And that is what Newton means by an unbalanced force in his second law.'

'What's Newton's third law?'

'*To every action there is an equal and opposite reaction.*'

'Oo!' Danny nearly put his hand up. 'I know that one.' He looked to the collision and smiled. 'That's why we can't say what happened in our crash, the effect is the same for each.'

'Exactly. The damage done by the Citroën hitting my Beetle is the same as the damage done by my Beetle hitting the Citroën because that collision, or *action* as Newton called it, produced one-size resultant force that acted on each body as though it had been hit by the other.'

'It seems obvious now.'

'All three laws become clear once you understand the principle of relative motion. Newton's first law deals with how a body will move if not acted on by an outside force, and says it will remain in uniform motion in a straight line. His second law deals with the effect of an outside force, which is to cause an acceleration of the body in the direction of the force, and the amount of acceleration is proportional to the size of the force and continues for as long as the force acts. And his third law says whenever bodies do collide each receives the same force acting in the opposite direction. These three simple laws describe how everything in our universe moves.'

61

'Everything?'

'Everything. We have only looked at two bodies moving along a straight line, but regardless of the angles at which they collide or whether they bounce off each other or crumple together, Newton's laws apply to all collisions between all bodies. And from a golf ball driven down the fairway to the mechanics of building an engine, all the movements and forces involved can be found using those laws.'

'That still doesn't tell us what happened in our crash?' said Danny, stuffing in a last half of sandwich.

'We can never know that from only our point of view in my Beetle.'

Danny's eyes started a bulging competition with his cheeks. 'Dagnabbit, Bert! I thought that was the point of doing the maths.'

'No. But do not worry, we still have my MFDP.' Bert collected their empty bottles into the empty sandwich box and headed for the Beetle. 'The program has nearly finished,' he called back, dropping into his seat. 'Now we will find out what happened from Spaceship Earth's point of view.'

6

Danny mooched around at the edge of the road, kicking everything from large stones to small French cars, as he checked for a signal on his mobile and tried to come up with a plan.

Armed with the fresh ammunition of a serious car crash, he was toying with a second assault on his mum. But even if she agreed – and it would have to include him sneaking off in a taxi leaving her to explain it to Bert – he'd only end up somewhere else. She didn't seem to care where.

He shoved the phone into his jeans and fell back against the Citroën.

The Rango Kid would do something. He'd jump into a railroad car and settle in the hay opposite a couple of sleeping hobos, chewing on a barley stalk as the endless plains clattered by. He'd rope a wild stallion, break it in a quiet river and ride bareback all the way to Phoenix. He'd steal the horse, rob a bank, shoot his way out of town, hole-up amongst a bunch of large rocks in the mountains and await the arrival of the posse. He'd –

As Bert emerged from the Beetle, his head bowed and his feet shuffling in the sand, a Little Bang of an idea went off in Danny's head and expanded into a universe of possibility. It was bold. It was reckless. It was crazy and would never work. 'It was us?'

Bert nodded but kept his head down.

Danny span round to re-survey the crash. 'The Citroën was standing still and we crashed into it at 30 mph?'

Bert nodded again.

'What about the emergency failsafe?'

'Like me, the computer confused the simulated collision with the real one. The emergency software was corrupted by the collision program, and vice versa, and they both crashed.' Bert glanced up. 'Thank goodness no one was hurt … it could have been much worse.

We were in spaceship mode and the program makes my Beetle act like she is in outer space.'

'Yeah. So?'

'So after the collision we kept going at 20 mph west and we took Morris' Citroën with us.'

'That's why there are no skid marks or broken glass.'

'We had only just rolled to a stop when the window screens cleared.'

Danny peered up the road for evidence glinting off the tarmac. 'Where did it happen?'

'Five and a half miles back.'

'Five and a half miles!'

'I know,' said Bert, twisting the toe of his boot into the sand.

'What about the driver?'

Bert raised his head. 'With us not stopping, and the empty weight of the Citroën fitting exactly with the measurements of what happened.' He paused and shrugged. 'He cannot have been in the car.'

'Where was he then?'

'Maybe he stopped for a pee.' Bert dropped his chin back to his chest and shuffled off to the Beetle. 'I had better call the Sheriff's department.'

'Oh Ma-an.' Danny's knees gave up the fight against gravity. His fledging plan required three things if it was ever going to fly, all of them – hopefully – at the Missing Horse. 'Bert?' he said, popping upright, like a meerkat, and looking both ways up the empty road. 'Why don't we … well … just go? It'll take hours if the police get involved.'

'I suppose we could leave it for Morris to sort out. That way we can get on with our expedition'

'Exactly. Anyway, how would you explain it all?'

Bert hmmed and tapped his finger to the side of his nose. 'Yes, it might be better to keep this in-house.'

'Go on,' said Danny, with a nudge to the old man's ribs.

And the hairs sticking out of Bert's ears all stood upright.

*

They prised open the Citroën's steel claws with a crowbar, chocked its rear wheels with large rocks, and after Danny told him all about G-K Formula One, Bert let him reverse the Beetle apart and bulldoze the Citroën sideways off the road.

Danny was still grinning as the Beetle accelerated west, following the line of the mid-afternoon sun towards the ridge and its early horizon.

'What if somebody saw us?' said Bert.

'On this road.' Danny watched the Citroën dissolve into a blob in the rear window then span back to the front. 'How much further is it to the Missing Horse?'

'The collision caused quite a delay and we have a way to go yet. But I am going to take a short cut. I do not want to meet Morris coming the other way.'

'Will he be mad?'

Bert snorted. 'He reverts to his native tongue when he is really angry. The only consolation is that the thief who stole his car got the pudding he deserved.'

Danny's face unwound as he realised what Bert was saying, and a better way of saying it. 'No,' he said, grabbing Bert's arm. 'He got an equal reaction to his action.'

'Ha, that's right,' said Bert, joining Danny in laughter but not the open-mouthed swaying in his seat. 'Which reminds me. Now we can see that the maths has revealed another deep truth about the universe.'

Danny swayed to a stop. 'Does *everything* have to be about science?'

'I thought you would want to know.'

'We-ll ... can't we do something fun?'

'Discovering universal truths is always fun. You see, the maths has been saying it is as though our collision never happened because, in a way, it did not happen.'

'In what way?'

'According to the maths, every body moving in an inertial frame is in its preferred state, which is equivalent to standing still with zero momentum. So the total momentum of all bodies before any collision is zero.'

'Then everything must be standing still with zero momentum after the collision as well.'

'Which means the total before and after must be the same. We can see this clearly if we look at our collision from Spaceship North Pole. You can check the maths for yourself and you will see that the total momentum was 2460 UM east, made up of the Citroën's 840 UM and my Beetle's 1620 UM before the collision. And the total momentum was still 2460 UM east but made up of the Citroën's 820 UM and my Beetle's 1640 UM after the collision.'

'Because each received an equal and opposite force.'

'Which cancelled out to leave no overall effect on the total momentum of the bodies as though the collision never happened. This is known as the Law of Conservation of Momentum, which simply states that *the total momentum of all bodies in a closed system remains the same regardless of the collisions that take place within the system.*'

'What's a closed system?'

'A closed system is one where the bodies within the system are not affected by anything outside the system. In our collision my Beetle and the Citroën were not a closed system because they were affected by the Earth's gravity and friction with its surface and atmosphere.'

'If they'd been spaceships in empty space, they would have been.'

'And any collision between them would have no overall effect on that system. The conservation of momentum also applies to single-body systems. According to Newton's first law, if a body is spinning then it wants to keep spinning at the same speed.'

'Doesn't it want to move in a straight line at the same speed?'

'Every part of it does, yes, but if everything is glued together in a solid body, like the Earth, then it cannot unless the speed of rotation and momentum become sufficiently great to tear it apart. But everything can keep moving at a constant speed, and in the vacuum of space where there is nothing to slow it down, this spinning inertia produces closed rotating systems like stars and planets, and the measurement we call *angular momentum*. Like linear or straight-line momentum, angular momentum is conserved, but the angular momentum of a spinning system depends on its mass, the distance or radius of that mass from the centre of the system, and the speed it is moving at that position.'

'You mean like, here, the Earth's surface is doing 840 mph but around the North Pole it's hardly moving?'

'In a way. The Earth is a rotating sphere with a fixed shape, which means the time to complete one rotation is the same for all points on its surface but the distance you must travel increases as you move further away from the axis of spin. This results in a surface speed rising from zero at the Poles to over 1000 mph at the Equator. But for a body that can change shape, like an ice skater spinning on the spot –'

'If they pull their arms in, they rotate faster.'

'When they extend their arms, they slow down again. As the distribution of their mass changes, their speed of rotation automatically alters by a proportional amount to conserve the angular momentum created by the initial spin. If there were no friction with the atmosphere and the ice, they would continue to rotate at a speed determined only by the radius they create with their body. But whether it is a spinning skater or planet, two spaceships or the entire universe, the maths has shown us that the total momentum is conserved.'

Bert twisted in his seat, his eyes big and shiny behind his glasses. 'That is why mathematics is called the language of Nature. The maths led us from mass times velocity equals zero for one body, to understanding forces, how everything in the universe moves under the action of those forces and how the momentum of that motion is conserved. And this is just the beginning. The conservation of momentum led to the conservation of energy.'

'Energy can't be created or destroyed.'

'That's right. We use energy to measure a body's ability to perform work, which means its ability to make things move.'

'That's what a force does.'

'No, in our collision the equal but opposite forces cancelled out. It was the energy of my moving Beetle prior to the collision that was delivered and caused the Citroen to move. This energy of motion is called *kinetic energy*, and the faster a body moves the more kinetic energy it has because it can cause another body to move further and so perform more work.'

Danny figured a moving car must deliver something, even if it wasn't a force. But then … 'Isn't that moving kinetic energy just the body's momentum?'

'In the late 1600's a German called Gottfried Leibniz performed a simple experiment that showed that it was not. Dropping a metal ball into soft clay, he discovered that when he doubled the speed of the ball —'

'Oh, I know this. Did it sink in four times as far? That's what happens to the braking distance of a car.' At G-K Formula One you had to watch a safe driving video before being allowed out in the karts to try and drift your mates off the track.

'Ha! There you go again,' beamed Bert. 'If you triple a car's speed, it takes nine times as far to stop. Quadruple the speed and it will take 16 times further and so on. In other words $E = mv^2$ – a body's velocity has to be squared to find its kinetic energy. And just as the total momentum of all motion in a system is conserved, so is the kinetic energy of that motion. If two identical cars travelling at the same speed collide from opposite directions, their equal and opposite forces cancel out, but their kinetic energy is converted into the heat energy of vibrating atoms in twisting metal and sliding rubber; the sound energy of vibrating air molecules; and the kinetic energy of flying debris. If we take them all into account we find that none of the original energy has been lost because one form of energy is always converted into another.'

Bert tapped the computer keypad and lines of code started scrolling down the screen. 'But kinetic energy is not the only form. When you hold a hammer above the ground it has the potential to start falling, which means gravity has energy because it can make bodies move and perform work. We call this energy of position *potential* or *gravitational energy* and it is also conserved.'

'Once the hammer's fallen to the ground hasn't it lost its gravitational energy?'

'Gravitational energy is called potential energy because it is really stored-up kinetic energy waiting to happen. In your hand the hammer starts off with the gravitational energy of its position, and the higher you raise it the more potential energy it has.'

'Because gravity has more time to accelerate it to a greater speed?'

'That's right. As it falls, it trades the decreasing potential energy of its lowering position for the increasing kinetic energy of its greater speed, until it hits the ground when all of its original gravitational energy is converted into its final speed. But if you think about it, that gravitational energy is simply being returned because it took the exact same amount of *your* kinetic energy to raise the hammer in the first place against the force of gravity that was constantly trying to pull it down. This means, mathematically, we can treat kinetic energy as being positive and gravitational energy negative. It could be the other way round, but as one grows the other shrinks by the same amount, always balancing to the total energy of the system.'

'Isn't that going to be zero?'

'Yes! Ha! I do not know why I bother. That is why we believe that the total energy of the universe is zero.'

'How can it be zero when everything's moving?'

'In the same way that it takes your kinetic energy to raise the hammer, in the expansion of the universe since the explosion of the Big Bang it has taken the same amount of positive kinetic energy to move everything apart and create all the negative gravitational energy that exists today. One cancels out the other and the total energy of the universe is zero.'

'Bert?' said Danny, turning to face him. 'Why do you bother?'

'Sorry?'

'It just seems weird.'

Bert started fiddling with his glasses. 'What seems weird?'

'Your science camp. Why go to all this effort for just one student?'

'Ah.' Bert raised his hands, as though Danny had pulled a gun. 'I admit I am cheating a little. My expedition will help you with your grades, far better than any school or college camp, but the truth is ... I need your help.'

'Can't your friend Morris help?'

'No-o. I need an unbiased, questioning mind, not someone who thinks they already know everything. Someone whose passion for cosmology will keep me on track. Their curiosity and desire to fully understand forcing me to re-examine what I think I know to be true as I try to solve the mystery ...' Bert opened his mouth, as though there was more, but nothing more came.

Boy, are you in for a disappointment, thought Danny. His only interest in physics now sat two rows down and one column to the left and wore really short skirts. 'How can I be any help when you know it all already?'

The ground climbed up the windscreen to an early horizon in more of a long, gradual slope than a ridge, and as the Beetle approached a T-junction and an entirely new road running along its base, Bert became lost in an horizon of his own.

'If only I did know it all,' he said, a sadness giving purpose to his watery old eyes and collapsing his body where all those years of gravity had failed.

To Danny, he suddenly looked very old and very tired.

'Something happened to me. Something that should not have happened and yet it did. It is imperative I find the answer ... but I do not know what it is or where it lies.'

'What happened, Bert?'

The computer beeped '**PLP test completed**' onto its screen, next to a traffic light with the bottom circle glowing bright green, and tossing gravity aside, Bert poked his pipe into Danny's ribs.

'You said you wanted to do something fun.' He slowed the Beetle and turned right at the T-junction, driving north along the base of the slope. 'Now that we understand Newton's three laws of motion, let's see why all bodies accelerate at g.'

'Oh yeah. Why does everything fall at the same rate?'

'Grab the crash helmets and I will show you.'

Danny pulled the helmets over his seat. 'Where did you get that?'

'From under my seat,' said Bert, a purple bowling ball lying heavy in his upturned hands.

'What are you going to do with it?'

'You mean what are *you* going to do with it.' Bert lobbed the bowling ball into Danny's lap, and as Danny's knees shot up to catch it between his stomach and thighs, he grabbed the steering wheel and swung it hard to the left.

Danny braced himself against Bert's seat, until his own started accelerating him up a narrow side road; its tarmac polished as though it had just stopped steaming and they'd drive over the top to find a giant resurfacing machine.

'This is the shortcut to the Missing Horse,' said Bert, trimming the opening to his helmet with a lace of white hair. 'Along the way we can do this experiment and ...' he threw Danny a grin, 'have some fun. Hold the bowling ball out of your window and when I say *now* let it fall.'

Danny loosened his seat belt and held out the ball, resting his elbows on the frame to help support its weight. 'What's dropping it from here going to prove?' he said, watching the brow of the hill fast approaching.

'Do not take your eyes off that ball, and let go when I say.'

As Danny's head snapped back to his charge, he noticed something – the word 'PANIC' had been scratched in inch-high letters around its middle. Rotating the ball he found 'YOU', after a gap 'WILL', closer together 'BE', a bit further on 'FINE' and then a 'DO'. The final 'NOT' did nothing to alter a growing feeling of unease. 'What's this mean?' he said, twisting to look at Bert but stopping halfway round.

'NOW!'

Like Shaggy and Scooby Doo when they first see the ghost, everything from Danny's hair down leapt backwards leaving only his eyeballs behind. Everything then proceeded to try and get into the back of the Beetle without going around or over the front seats, and everything became desperately confused when it found that it couldn't. A deeper, older and more powerful area of Danny's brain had taken over, locking his body into flight mode, but with nowhere to flee, locking his eyes on the windscreen and the horror that lay ahead.

It wasn't a hill it was a cliff, and it wasn't so much a cliff as a canyon. It was a 4000-foot drop at the eastern end of the main Grand Canyon wall and Bert had driven straight off the top.

z

The Beetle soared off the natural ramp formed by the slope and the edge of the Canyon; and with his mouth open wide, his skin stretched tight and his helmet vibrating on his head, like an alarm bell, Danny started to scream.

'Aaaaagh-urrrgh!'

With waving arms grabbing for a hold, he floated into the air, bounced against the loose straps of his seat belt and hovered above his seat. Then he landed again, quickly followed by his stomach.

'Aaaaaaaaaagggh!'

He shot Bert an unfocused look of disbelief and shared distress as the Beetle tilted into a dive and a distant mountainous wall emerged from the bottom of the windscreen, squashing the sky into a diminishing stripe of blue.

'Aaaaaaaaaagggh!'

A Martian-red landscape of jagged valleys and winding gorges followed, slowing as the Beetle homed in on one of its staggering canyons, and growing as it hurtled between its vertiginous walls. Vertical bands of rock flashed across the side windows, but as Danny gasped, to fuel another scream, it occurred to him that his distress was not being shared as much as he felt it should.

'Do not panic,' said Bert, squashed in beside the steering wheel with his back against the door, his feet up on the seat and a camera pressed to his eye. 'You will be fine.'

'WHATTHE HELLAREYOUDOING!'

'Can you move back a bit?' said Bert, waving a hand in front of the camera to shoo him out of the way.

Danny's helmet fell forward as the Beetle rocked violently twice. But when the Grim Reaper's death grip turned out to be an uncomfortable pressure across his stomach and chest, he pushed up the helmet and forced open slits in the squeeze of his eyes.

'Ahwaaha-huh?'

'That was the two main parachutes,' replied Bert.

The canyon floor spiralled towards the windscreen, with Danny suspended in his seat belt above it. Then something fired up the front of the Beetle and a fan of parachute cords snapped open, jerking them to the angle of a child's slide and slowing their approach to a brisk walk.

'That was the third.'

The Beetle pulled out of its spiral, glided through the canyon, and floated down onto her crash bars and front wheels with nothing like the force of their previous collision. As she bounced down onto her back wheels with even less, a shroud of parachute silk settled over her, blocking out the windows and filtering the sunbeams into soft, yellow light.

'Was that enough fun for you?' said Bert, peering down, like Old Mother Hubbard in her bonnet.

Danny rolled his helmet along his seat. 'Wow,' he said, puffing out an exhausted smile. He patted parts of the Beetle and his body, as though checking they were all there, then jumped up and grabbed Bert's arm. 'Can we do it again?'

'Ha! You are lucky we did it at all.'

'I've got to see where we've landed.'

'Go ahead. I am going to re-run the MFDP.'

Danny pressed a mould of his door into the silk, followed by a mould of his helmet, and crawled underneath the parachute.

The Beetle had landed in a deep canyon – narrowly missing a gnarled juniper tree – on the sweeping bend of an intermittent spur to the Colorado River that was now a ribbon of sand. Rock walls soared thousands of feet on both sides, first in chiselled steps, like crumbling Mayan temples, then autumn-red cliffs, curtaining the sky towards the head of the canyon. There they met the main Canyon wall rising thousands of feet higher still in layer upon layer of red and yellow with a dazzling icing of white.

Springing to his feet, Danny started jumping up and down, his fists punching the sky as his whoops echoed through the canyon.

'Are you okay?' called a deep voice.

Danny span round to find three boys in t-shirts, twisted baseball caps and falling jeans gawping at him from 40 feet away.

'We were on our way home.'

It was the tall thin boy, standing as though he'd been ambushed by his body as well as his voice.

'Hey, guys!' said Danny, throwing off his helmet and bouncing towards them, his arm pointing back to the wall. 'Did you see that? Whoo-wee! It was meana' than a bronc in a corral full a' firecrackers.' He looked from face to face with a grin that lasted too long and cried out for some form of sedation.

'Man, it was amazing!' said the stocky, ginger-haired boy, pushing out his hands and breaking at the knees, as though performing a four-word rap.

From his face Danny reckoned puberty was having a go at him as well. The third boy, who was only about eight, just stood there staring at him.

'Where you from?' said Ginger. 'Texas?'

Danny wondered which way to go.

'You weren't driving that,' cut in Gangly, with the cocky assuredness of an extra eight inches of height and a two-year gap in age.

Danny glared up at him as the statement settled in his brain. 'I'm outta Tombstone,' he said, rocking his weight onto his right leg and looking back at Ginger. 'Right 'ere in the good ol' state a' Arizona.'

'You're not old enough to drive,' said Gangly, puffing up his chest, but glancing at his friend.

Danny pushed his hands deep into his pockets and turned for a show-the-most-Y-front face-off with the boy. 'Listen kid,' he said, his eyes cold between strands of matted hair, like the Rango Kid before a duel. 'I'm a dang sight older than you. I'm twenty-two but got me a rare brain disease is all. Stopped growin' when I was fourteen, not much younga' 'an you.' He spat a flake of imaginary tobacco off to the side. 'Now what d'ya boys mean, you were on your way –'

'What's the disease called?'

'Mishkinson's Thrombosis,' Danny cobbled it together. 'Look it up.'

As he was about to turn away, Gangly pulled out a mobile and started tapping on its screen. He held the phone high, followed it to either side of his head and dropped it back in his pocket. 'I meant we were passing in my car.'

'It's his mom's car,' said Ginger, receiving a look in return.

'What car?' said Danny, his eyes sweeping the canyon floor, like lighthouse beams. A few hundred yards downstream, the canyon walls overlapped and a blue, Matchbox-sized pickup sat alone at their intersection, as though it too had dropped in from nowhere. 'Is that a road?'

'It's an old track. It leads down to the ferry crossing and the highway,' said Gangly, lowering his eyes from Danny's.

'His mom's got a motel there,' said Ginger.

Gangly's head snapped round for a full-blown glare, but Ginger's shoulders bobbed and his hands twisted outward at his sides. 'I caught your flight on my camera,' he said, switching back to Danny and smiling as though that had been his job. 'Man, you should see it. We thought you were going to die.'

'Where is it?'

'In Joe's mom's car.'

As one the boys looked to the Beetle, and Danny turned to see Bert crawl out from underneath a parachute and start peeling it off the bonnet.

'Who's that?' said Joe.

'Oh ... er ... that there's m'assistant,' said Danny, wandering casually in the direction of their truck. 'Now you fellas can see why I was driving. The name's Danny. Danny Rango – call me Rango.' He stopped and let each boy come forward to shake his outstretched hand; Joe, then Ginger a.k.a. Billy, but the young kid was obviously too stricken with awe.

'That's my brother, Christopher,' said Billy. 'He doesn't talk much.'

'I have done a systems check,' called Bert from the back of the Beetle. 'I am afraid we have sustained some damage.'

'Gimme a minute there boys.' Danny raised a hand of acknowledgement to Bert. 'Good,' he shouted, as he started strolling backwards away from him. 'Well done ... um ... keep up the good work an' I'll be back faster than a lead plumb from a peacemaker.'

Turning in mid-stride, he linked his hands behind his back and marched towards the pickup with the other boys running to catch up.

Skipping along beside him, Christopher finally plucked up the nerve. 'Rango?' he said, looking up as though he were looking down.

'Yeah,' said Danny, smiling encouragingly.

'Our grandma's got thrombosis in her leg.'

*

Danny stayed with the boys for half an hour, sitting on the Chevrolet's rusting tailgate as they jostled for position around him and watched the flight of the Beetle over and over again – the camera swinging to lock onto the tiny bug in the top left corner of the screen, accompanied by gasps from the boys on the sound track and pick up; the shocked silence of its dive to oblivion; wild cheering as the parachutes burst open, white angel wings against a heavenly blue sky; and the spiral descent and leaping applause on landing.

After soaking-up the boys' pats on the back, and acknowledging the size of his cojones with a sideways nod, Danny exchanged numbers with his new best-buddy, Joe, and with a brochure for *Rosie's*, his mom's motel, swinging in his hand, he meandered back across the sun-baked river of sand to discover that Bert had been busy.

A parachute canopy billowed between the Beetle's roof rack on one side, and the juniper tree and a guy-roped telescopic pole on the other. Shaded beneath sat the metal-framed chairs and blue cool-box, a pine table, the open briefcase, a towel covered with odd-shaped lumps of rock, a few inches of hairless white shins, and two feet of skyward-pointing hobnailed boot.

'Hi Bert,' he said, giving one of the boots a kick that sent them both scuttling under the Beetle.

Bert popped up on the far side of the bonnet, his forehead and cheeks striped with oil and his hair sprinkled with grains of earth, like chocolate on top of a cappuccino. 'Danny! You are back!' he said, skipping around the crash bars. 'This calls for a drink.' He pulled a bottle of cola and an *Apache* beer from the cool-box and popped their caps on the edge of the table.

76

Danny clinked his bottle into Bert's, took a long swallow of ice-cold cola, and coughed half of it back up. 'Those boys filmed the whole flight,' he said, wiping his mouth and dropping onto a chair. 'You should see it, Bert. They're going to send it to me.'

'What did you tell them?'

'I lied,' said Danny, and he was beginning to think that he may have overdone it. Intoxicated by their eager faces and a feeling of mischievous light-headedness, the final story had ended up: Danny Rango, a maths genius since the onset of his brain disease – this positive side-effect balanced by his probable death before the age of twenty-five – recruited by NASA to perform secret research experiments into gravity; a knot of controlled fear in his stomach, busily checking systems and firing parachutes whilst trying to calm his nervous assistant who wasn't his granddad. 'I told them we were scientists.'

'That is not lying. We are performing experiments to explore Nature's deepest mysteries and you cannot get more scientific than that.'

'Yeah, what was the point of that experiment?' Luckily the boys hadn't asked any probing questions. 'I thought you were going to show me why everything falls at the same rate?'

'Did you not see?'

Bert pressed a key on the briefcase computer and a flickering Danny settled in the middle of the screen, his head facing the front of the Beetle and his body twisted towards the back.

'This is my recording from inside the Beetle,' said Bert, pulling a penknife from his pocket and selecting a rock from the towel. 'We have just gone off the edge and I have … turned the sound down.'

Danny cringed as he watched himself vibrating with silent effort through the hole in his crash helmet – and Bert's shoulders bobbing up and down behind the briefcase lid. But by the time his body on screen had returned to its seat, and his face had gone from panic to surprise and back again, he was swaying in his chair with his mouth hanging open. Then a large, unfocused hand flapped briefly across the frame and the camera shifted to the left.

'It's the bowling ball!' cried Danny, his arm shooting out to the screen as a purple sun rose from behind his body.

The ball hovered outside the passenger window, as though caught in a fountain of air, and then disappeared below it as the picture jolted violently twice.

'The two main parachutes have just deployed from the metal cylinders at the rear of the roof rack,' said Bert, attacking the rock with his knife.

'The dustbins?' Danny realised he was hanging in his seat belt, but because Bert had kept the camera fixed in the same position throughout the flight, on screen it looked like he was sitting in his seat but trying desperately to pull himself out of it.

'The third parachute came from underneath the crash bars. They were fired by the PLP, my Parachute Landing Program, as the Beetle's altimeters registered each programmed height, and working together they formed a steerable wing that enabled us land in the middle of this canyon.'

'Then all we've done is repeat what the astronauts did on the Moon.'

'Yes, but we are not on the Moon.' Bert restarted the video. 'Keep watching ... there! Can you see?'

As Danny levitated above his seat in the background, Bert removed his hand in the foreground to leave his pipe invisibly suspended and revolving in mid-air, like Tom Hanks' glove in *Apollo Thirteen*.

Danny laughed as the pipe bounced off the dashboard and span slowly out of frame. 'How can that happen?'

'You tell me. What did it feel like?'

'I felt weightless.'

'How could you be weightless if weight is a measure of the amount of matter or stuff in your body and all that matter was still there? In other words, what are we actually measuring when we measure a body's weight? According to Newton's second law of motion, what is the effect of a force on a body?'

'It makes it accelerate in the direction of the force.'

'What is the effect of a continuous force?'

'A continuous acceleration at a constant rate.'

'Then what would happen, right now, if the ground opened beneath our feet?'

'We'd start falling.'

'Which means gravity does not disappear as soon as you hit the ground or if you are supported in mid-air in some way; it is *always* acting on you and it is *always* trying to make you fall.'

'You mean my weight is gravity forcing me into the ground?'

'Exactly. And the only reason you are not accelerating down under that continuous force, right now, is because the ground is in the way exerting an equal and opposite force in the other direction.'

'The two forces cancel out to leave me standing still.'

'So Newton introduced the scalar measurement of mass to describe the amount of matter from which you are made, as distinct to your weight, which is a vector measurement of the force of gravity acting on that mass. Here on Earth you weigh 50 kg so we make this the standard measurement and say your mass is 50 kg. But if you stood on your bathroom scales on the surface of the Moon, where the force of gravity is only a sixth that on Earth, you would still have a mass of 50 kg but you would weigh less than your rucksack.'

'That's why the astronauts can make giant leaps.'

'That's right,' said Bert, blowing a cloud of dust off his rock, which was now long and thin with a round bit at one end. 'Weight is a measure of force, and a *newton* – the force needed to accelerate one kilogram of mass by one metre per second every second – is equivalent to 100 grams or about the weight of an apple. So your 50 kg mass is equivalent to 500 newtons of gravitational force, or weight, pulling you into the ground.'

'How does that explain why everything falls at g?'

'What would you weigh if you ate enough to double your mass?'

'100 kg.'

'And because your weight is a measure of the force of gravity ...'

'The gravity must have doubled.'

'Which means the size or magnitude of the force of gravity acting on a body is directly proportional to the mass of that body. You have 500 times the weight of a 100-gram apple because you have 500 times more mass and your body is being subjected to a 500 times stronger gravitational force.'

'But if I double my mass and the force of gravity doubles, then according to Newton's second law I should accelerate twice as fast and nothing should fall at the same rate.'

'You already know the answer. If we were to have a pushing race between my Beetle and that massive lorry that passed earlier, which would you choose?'

'Your Beetle.' Danny smiled. 'If it's a bigger mass the force has less effect.'

'Exactly. Newton's second law says that the acceleration is directly proportional to the size of the force, which means it increases in the same proportion as any increase in force. But because every body possesses inertia, the acceleration is *inversely proportional* to the mass being accelerated, which means it *decreases* in the same proportion as any increase in mass. So if we drop my Beetle and Morris' Citroen, and let's say the lorry has a mass of 40 tonnes and we drop that as well, all from the exact same height at the exact same time, what happens? They are standing still relative to each other when we let them go, and the gravitational force acting on my Beetle is twice as strong as the force acting on the Citroen and forty times greater for the lorry.'

'But your Beetle has twice the inertia resisting the force and the lorry has forty times the inertia, one cancels out the other and everything falls at the same rate.' Danny was surprised at how simple it was, but he was more surprised to discover that Bert had whittled his rock into some sort of metal rod that appeared to belong underneath the Beetle.

Bert selected another lump from the towel. 'Any increase in the mass of a body increases the size of the force of gravity by the same amount that the body's increased inertia lessens its effect, leaving g, one constant rate of fall for all bodies.'

'I thought the atmosphere stopped that happening.'

'It does. Friction with the atmosphere is the result of billions of tiny collisions with billions of tiny air molecules. A greater mass has more momentum so the force of friction has less effect and it falls faster than a body with less mass but of the same size and shape.'

'Then why did the bowling ball fell at the same rate we did?'

'Because my Beetle and the ball are not the same size and shape. My Beetle's greater momentum compared to the bowling ball was cancelled by her greater size and surface area colliding with more molecules compared to the ball, and they both fell at the same rate, even with the atmosphere. That is not the only factor at work. Because all movement is relative, when you stick your head out of a car window driving at 70 mph on a still day, you know what it is like to stand still in a 70-mph wind. In other words, the faster you are moving through the atmosphere, the faster all those molecules of air are moving towards you and the greater *their* momentum.'

'Friction increases with speed as well.'

'Which is why, here on Earth, if you drop a body from high enough, as it accelerates down friction with the atmosphere builds up until the point where it is equal in strength to the force of gravity but pushing in the opposite direction. The two forces cancel out and the body stops accelerating and falls at a constant speed we call its terminal velocity.'

'Like skydivers.'

'It happens to all falling bodies given enough time, but each body's terminal velocity will depend on its mass and surface area. Skydivers reach speeds approaching 120 mph to achieve terminal velocity, whereas a feather reaches it as soon as it starts to fall.'

'If nothing got in the way, where would everything stop?'

'What happens when you hang a weight from a string?'

'It hangs straight down so that you can tell if something's vertical.'

'It hangs perpendicular or at 90 degrees to the ground in all directions at that point. And because the Earth is a huge ball, or sphere, this tells us that the direction of the force of gravity everywhere on its surface must be towards its very centre.'

'You mean everything's trying to fall to the same point?'

'That's right. We can imagine hanging millions of plumb lines over the planet surface, and if we extend their lines of fall, like a giant spiny urchin, they would all hit the centre of the Earth. We call this point where the lines of force meet the Earth's *centre of gravity* or *centre of mass* and that is where everything would stop.'

Danny pulled a bottle of cola and an Apache beer from the cool-box, smiling at the Indian Chief pictured on the label. 'If my weight is gravity and it's always acting on me,' he said, passing the beer to Bert, 'why was I weightless in your Beetle?'

'You were in orbit.'

'In orbit!'

Bert lifted his head from his work and grinned. 'To see why, we will have to measure the hidden force of gravity. But now that we understand Newton's three laws of motion everything is falling into place; and this is just the beginning! What else can we work out?'

$\mathcal{8}$

The shadow of the canyon's western walls crept across the sand towards the front of the Beetle, and as the occasional breeze inflated the parachute above their heads, Danny watched a thick spring materialise beneath Bert's penknife.

'We can now work out the extent of the force of gravity,' said Bert. 'According to Newton's second law, it takes the interaction of two bodies to produce a force.'

'Shouldn't they have to collide to feel the force?'

'It is true that in the mid-seventeenth century gravity is a mysterious force acting between bodies over great distances, like an invisible piece of string, but it is still a force and it still needs the interaction of both bodies. What does Newton's third law say?'

'To every action there is an equal and opposite reaction.'

'Which means both bodies dictate the strength of the gravitational force and the force pulling an apple towards the ground must also act on the ground pulling it towards the apple. The force has only a tiny effect on the Earth compared to its effect on the apple, but we know from Newton's third law that the effect must be there. That in turn means it is not just the sun and the planets that possess this mysterious force of attraction, every body does. And because we know that the Earth's gravity continuously pulls everything towards its centre of mass, we know that every body must have a centre of gravity as well.'

Danny fingered his belly button. 'Is it always right in the middle?'

'For a sphere or a cube it is the exact centre, but provided all of a body's mass is evenly distributed around a point, the body will balance on that point because gravity and therefore all its weight is acting through that point.'

'Oh yeah. My dad had a plastic eagle that could bob and spin around on a little plinth, balancing on its beak, even though the rest of its body was hanging out in space.' Danny knew it had something to do with its swept-forward wings, heavily weighted at the tips, but he'd never understood why. 'I know how to do it with a match and two forks.'

'The centre of gravity can be outside the body altogether. It is somewhere in the space inside a cup, and a high-jumper's passes under the bar even though their entire body passes over it.'

'Because of the way they bend.'

'The point is every body attracts every other body towards its centre of gravity because *gravity is a property of mass*. Sitting there now, there is a tiny resultant gravitational force acting between you and your bottle, you and the table, and you and me, continuously pulling us towards each other.'

Danny could feel the bottle's and it was drawn to his lips as proof.

'And because the gravity between two bodies is the resultant force from the mutual attraction of both masses, just as doubling your mass doubles the force acting on you, if the mass of the Earth doubled instead ...'

'I'd weigh 100 kg again.'

'If your mass doubled and the mass of the Earth tripled?'

'Would it be 300 kg?'

'That's right. We have to multiply the bodies' masses to find the strength of the force of gravity acting between them and you would weigh six times more. If you could stand on the surface of Jupiter, the product of your masses would produce a resultant gravitational force two-and-a-half times greater than you experience at the Earth's surface and you would weigh 125 kg. Whereas you weigh only 8.3 kg on the Moon's surface because the resultant force is only a sixth.'

'I thought Jupiter was huge.'

'Jupiter is over a thousand times larger than the Earth, but it is made mostly of gas and has only 318 times more actual mass.'

'That still isn't right. Why wouldn't I weigh ... 318 times more?'

'Ah, that is a good question,' said Bert, wiping an L-shaped bracket with his shirt and placing it on the towel. 'You see, it had always been believed that the Moon and the planets move in circular orbits at a constant speed, but in 1609 a German astronomer called

Johannes Kepler discovered that their orbits are elliptical. An Ellipse is a conic section.'

'Like a parabola.'

'Yes, except it is made by slicing all the way through a cone as oppose to dissecting its base. If you make a horizontal slice, you get a conic section called a circle mathematically constructed around one focal point, its centre. If you slice at an angle to the base, you get a conic section called an ellipse, which is like an oval or stretched circle with the stretching increasing in the direction of the cone tip as you increase the angle of the slice. An ellipse is constructed around two focal points, and Kepler discovered that the planets move in elliptical orbits with the sun at the smaller, stretched focal point of each ellipse. He also discovered that the further a planet is from the sun, the slower it moves in its elliptical orbit, and its orbital velocity is not constant. Like two giant parabolas joined together, it slows down as it moves away from the sun and accelerates as it approaches.'

Bert gave Danny one of his now-we-can-see grins, and Danny sent an oh-no-we-can't frown back.

'Newton realised that if a planet is moving more slowly it means it is experiencing less acceleration, and under his second law, if a body experiences less acceleration ...'

'It must be a weaker force.'

'The force of gravity must weaken with distance. But what is the mathematical relationship?'

Danny puffed out a mouthful of air.

'When you were twirling Io and Ganymede, at what point could they resume moving in a straight line?'

'Whenever I let go of the string.'

'Which means at every instant in their orbit the inward force of you holding the string balanced the centrifugal effect of the ball's inertia trying to return it to a straight line at that instant. The two effects were equal but acting in opposite directions and –'

'Cancelled each other out.'

'Newton realised that if he could measure the Moon's momentum at different points in its elliptical orbit, that would give him a measure of the force of gravity preventing it from returning to a straight line at those points. After discovering a new branch of mathematics called *Calculus*, which enabled him to measure the velocity of an object moving on a curve, he found that the strength of the resultant gravitational force between two bodies is inversely proportional to the square of the distance between their centres of gravity. It sounds horrible but simply means, if you double the distance the gravitational force will be four times weaker; if you triple the distance it will be nine times weaker; and if you quadruple it –'

'It'll be 16 times weaker.'

'And so on,' said Bert, sculpting a winding thread into what was destined to become a thick bolt. 'So if Jupiter's mass were squeezed into the volume of planet Earth, you would be 318 times heavier because you would be the same distance from the centre of all that mass as you are, sitting here now, from the Earth's centre. But because Jupiter is so large, its surface is over 12 times further away, its surface gravity is 144 times weaker and you only weigh 125 kg.'

Bert swept up his bottle and clinked it against Danny's. 'We have done it!' he beamed. 'That is Newton's Universal Law of Gravitation, which simply states that *every body attracts every other body towards its centre of mass with a force that is directly proportional to the product of both masses and inversely proportional to the square of the distance between them.*'

'Is that it?'

'Now we can see why you were weightless in my Beetle. And it is the same reason the astronauts are weightless as they orbit in the International Space Station.'

'But there's zero gravity in space so how can it be the same?'

'Zero gravity does not mean the force disappears when you venture into space. The continuous force of gravity that gives bodies their weight down here on Earth also keeps the International Space Station in orbit and reaches out a further 250,000 miles to the Moon.'

'Yeah, but on Earth everything accelerates straight down towards its centre, not round and round.'

'Bodies on Earth do not always fall straight down. What happens when you fire a bullet from a gun?'

'It does a parabola.'

'Newton realised, that is what happens in an orbit. The Moon's momentum tries to keep it moving in a straight line in the vacuum of space, but it is constantly falling at *g*, the rate of fall for all bodies, as the Earth pulls it towards its centre of gravity. The combination of these two effects produces a parabolic curve towards the ground, the only difference is the Moon misses.'

'Eh?'

'Let's look at it the other way round and put a bullet into orbit. When a bullet is fired from a gun, level to the ground, it travels a long way before gravity has the chance to accelerate it down by much at all, so the first part of its parabola is a long, slight, downward curve. Now imagine we have climbed to the top of Mount Everest with a Supergun; a gun that can fire a bullet so fast and cover so much ground in such a short time, its parabolic flight-path follows the curvature of the Earth.'

'Oh-h, as it curves towards the ground, the ground curves away beneath it.'

'Exactly. And if there were no friction with the atmosphere –'

'It would keep going round in orbit and never get closer to the ground.'

'That is what is happening to the International Space Station and the Moon, right now, and for that brief period you were weightless, that is what happened to you. As soon as we left the top of the Canyon, my Beetle and everything inside started to fall.'

'But we kept going up.'

'Our momentum kept us rising for a while, but we did not keep moving up in a straight line. We had started to fall at g into a parabola, and because your body and my pipe were falling at the same constant rate as the Beetle, nothing got in their way to exert a force in the other direction and they floated weightlessly in mid air. Until friction with the atmosphere caused the Beetle to fall more slowly than everything inside and you caught up with your seat and my pipe with the foot well.'

'I could feel my weight again.'

'Not your full weight. You became progressively heavier as the Beetle's speed increased and the force of friction slowed her rate of fall even more.'

'I did when the parachutes opened.'

'Yes, because from that point on the Beetle was falling at a constant velocity, as though she had reached terminal velocity or was simply sitting on the ground. But for an astronaut on the International Space Station, orbiting the Earth in the vacuum of space, nothing ever gets in the way. He is always falling at g, and because the station is falling around him at the same rate along the same elliptical path, and at every instant his straight-line inertia is cancelled by gravity and vice versa, he does not drift up to the ceiling and he is not pulled to the floor, but remains locked in a continuous orbit by a resultant force he cannot measure from his point of view.'

'So zero gravity means you can't feel its effect.'

'And that is Newton's theory of gravity.'

Bert smiled at Danny relaxing in his chair. 'It is now 1687,' he said, picking up the last fossilised car-part. 'After lying hidden in his notebooks for over 20 years, Newton publishes his findings and lays the foundation for our modern technological age. His three laws of motion explain how all bodies move under the action of forces, and his force equation, $F = ma$, shows us how to manipulate bodies by the controlled use of forces. This forms the basis of a new science of motion called Mechanics, which revolutionises the world by giving us machines.'

'Why did he wait 20 years?'

'He was a brilliant but difficult man and did not like criticism. Who can blame him when his law of gravitation explains why everything falls here on Earth; how the Moon's gravity pulls on the oceans to produce the tides; and why moons orbit planets, planets orbit suns and the deviations caused by other planets passing nearby. It was a wobble in the orbit of Uranus that led to the discovery of Neptune in 1846. Then in 1957 the Russians combined all four laws and launched Sputnik One, the first manmade satellite. For three months the 23.5-inch diameter aluminium ball with four trailing antennae orbited the Earth, like a silver shuttle cock, beeping down the first-ever message from space.'

'What was the message?'

'Beep … Beep … Beep … But though it started as a Cold War propaganda exercise, its ultimate message was loud and clear and full of hope: our exploration of space had begun.'

'Why was it only in orbit for three months?'

'For a satellite to maintain a closed orbit around the Earth it has to be high enough to avoid contact with the atmosphere and be moving with sufficient momentum to maintain that height. The closest approach of Sputnik One's elliptical orbit was only 142 miles above the Earth, and though the air at that height is incredibly thin it still comprises billions of molecules. Every resultant force from every tiny collision acted to slow the satellite down, dropping her into the denser air of a lower orbit and slowing her even more, until she burned to nothing in a streak of light to become the first manmade shooting star as well.'

'How high do you have to go to stay in orbit?'

'The Space Shuttle used to operate at around 200 miles and at 18,000 mph, depending on the ellipse of its orbit. The International Space Station orbits at around 250 miles at an average 17,500 mph.'

'It moves more slowly because it's further out in space and the gravity's weaker?'

'That's right. The Moon takes over 27 days to complete an orbit at an average 2200 mph. And if you place a satellite into a geostationary orbit, moving west to east at a height of 22,300 miles above the equator, its orbital period will be 24 hours and it will rotate with the same angular velocity as the Earth and maintain the same position relative to the ground.'

'I always wondered what that meant.'

Bert lobbed a sturdy metal ring onto the towel and lifted its corners into a bag of bits. 'Isn't it amazing? This morning we discovered that all movement is relative, and now we understand the conservation of momentum and energy and the motion of the entire universe under Newton's theory of gravity.'

'What are those?' said Danny, nodding to the towel.

'They form part of the Beetle's steering rack and front-left suspension,' said Bert, dropping to his knees and rolling onto his back. 'It is very interesting as it operates according to the Ackerman geometric principle that the inside wheel –'

'How long will it take to fix?'

*

It took Bert over an hour-and-a-half to rebuild the Beetle's suspension, and Danny spent most of it wandering up the canyon in a fruitless search for the bowling ball. When the ground became too steep, and a sharp shadow-line started climbing up his boots, he watched the sun dip below the canyon's overlapping western walls and worked his way back to the Beetle.

'You are back!' said Bert, pulling his fingers off the keypad and looking up from the driver's seat. 'So soon, ha-ha-ha.'

'What's wrong?'

'No-o. No, it is nothing I cannot – that is I should be able to – I mean –' Bert's bravado collapsed with his face. 'She won't start.' He pointed to the mass of numbers filling the oval screen. 'The MFDP says there is now an electrical fault.'

'Oh.'

'I thought you would be upset.'

'Upset?'

'I am trying to locate the problem, but it goes dark early down here and I don't see too well at night, so … well … it may be too late to make the Missing Horse.' Bert glanced up and hurried on. 'It would be best to camp here and make the final push in the morning. Hey! But if you like, we can get my telescope out later. I have made some improvements and have been meaning to try them out.'

Bert offered up the telescope as though it would make everything better. And as the old man grinned up at him, like a troll on the end of a pencil, Danny was surprised to discover that it sort of did.

'Unfortunately for you,' said Bert, 'I had already packed away our camp.'

'For me?' Danny smiled and started loosening the ratchet strap above the window.

'You will find wood for a fire under the tarpaulin at the front of the roof rack,' said Bert, his fingers already back at work.

Danny jumped onto the Beetle's crash bars and started throwing logs onto the sand. 'What's in all these boxes?'

'Equipment and supplies for our expedition, plus the odd experiment.' Bert stuck his head out of the window. 'That reminds me. While you are up there, bring down the crate marked *BJ*. It should be behind the firewood, next to the large white chest.'

Bert worked on the problem for as long as it took Danny to

unload most of the camp and set up the table and chairs. Twenty minutes later, a fire crackled next the juniper tree and the old kettle hanging over it from a tent-frame of sticks was just beginning to steam. A second, smaller table supported a plastic bowl and a towel, and a lantern hanging from the roof rack combined with the fire to give everything a soft yellow glow in the gathering dusk.

'By the way,' said Bert, emptying the kettle into the bowl. 'We were meant to be celebrating this at the Missing Horse, but I thought you might like to know that we have completed Newton and stage one of our expedition.'

'I forgot this,' said Danny, sliding off the Beetle's bonnet with the wooden BJ crate in his arms. He dumped it by the front wheel and winced as a delicate ping reverberated from inside. 'What's going to happen on stage two?'

'First, we will find out why Newton's theories are incorrect.'

Danny gawped at Bert. 'Newton's wrong?'

'His theories are right as far as they go, and they go a long way; we used them to land men on the Moon and we will use them to colonise Mars. But they only explain how gravity works, not what causes it.'

'Mass.'

'But why is gravity a property of mass? How does it reach invisibly across space to pull bodies together? And according to Newton's law of gravitation, if the distance between two bodies increases by a factor of ten –'

'The gravity between them will be 100 times weaker.'

'If it increases by a factor of 100, the force will be 10,000 times weaker and so on. This means there will always be a tiny gravitational force acting between bodies, regardless of their distance apart, so why isn't everything pulled into one big lump in the middle of the universe?'

'Is it because the universe is expanding?'

'That's right. But seventeenth-century technology could not detect the relative motion of stars, let alone galaxies, and the universe was believed to be standing still. In any event, until we find the *Theory of Everything* all theories will hit obstacles they cannot explain their way around. It would take 200 years to get there, but Newton was heading straight for Mount Everest and we must follow in Einstein's footsteps to see why.'

'Einstein? We don't study him.'

'Really? How are you expected to appreciate and develop an interest in physics if you do not look at its most exciting achievements? And a proper understanding of Newton is vital to understanding Einstein because he made the next leap. His theory of relativity explains *what* gravity is and *why* it exists as a property of mass. It shows that the principle of relativity goes far deeper than we ever suspected, and like all measurements of motion, measurements of space and time are also relative.'

'Bert,' said Danny, dropping on to a chair. 'When are we going to eat? I'm starving.'

9

They started a meal of roasted chicken, baked potatoes and pinto beans with an orange wash crossing the sky from the west, but by the time they'd finished their plates – and reliving the Beetle's flight into the Canyon – their campfire was the dominant source of light.

Danny interlocked the tines of their two forks and pressed them smoothly together. Using Bert's penknife to sharpen one end of a toothpick-sized twig, he forced it between the tines and rested the other end on the top of his cola bottle. 'Da-daa!' he said, as the forks balanced impossibly on one side, like his dad's plastic eagle balancing on its beak. He touched the end of one of the forks and the contraption rocked slowly up and down.

'So tell me,' said Bert, the firelight dancing in his glasses. 'Why is such a smart young man failing his physics exams?'

Danny's finger tapped the forks too hard and they clattered onto the table. 'I don't know,' he said, picking up the bottle and sinking back into his chair. 'It's so … boring, I just lost interest.'

'The boy who did that project did not find it boring. And was it boring today?'

Danny said nothing.

'It is not surprising. Teaching should be such that students perceive it as a valuable gift and not a hard duty, but it seldom is and too many childhood interests are crushed by a formal education; it nearly happened to me. What fired your early interest?'

Danny's mind flashed back to the excited nine-year-old leaning over his dad's chair as Neil Armstrong watched from the surface of the Moon above the desk and his dad turned the large blue pages of his project book.

'My dad was an engineer and really into space,' he said, staring into the opening of his bottle. 'He built a clockwork model of the

93

solar system and had loads of books on science and photos of the Apollo missions.' He glanced up and smiled. 'He even had one of you.'

Bert sucked quietly on his pipe and nodded, as though that were a perfectly normal thing to possess. Then the pipe stood to attention. 'Oh,' he said, ducking his head into his shoulders. 'You mean Einstein.'

'He had all sorts of science toys, like the balancing eagle. My favourite was a row of six metal balls hanging on strings, and however many you let swing into one end of the row, the same number swing away at the other end and it keeps going for ages.'

'Newton's Cradle. I like that one.'

Danny's forehead creased beneath his fringe. 'Does it work because the momentum and energy of the balls is conserved?'

'That's right. The energy of the swinging ball travels through the middle balls as a wave and is passed on to the end ball. If there were no loss of heat energy due to friction with the strings and the atmosphere, it would keep passing that energy back and forth and never stop. Do you still play with it?'

'We couldn't afford to stay in the old house and it all got put into storage.'

Danny lowered his head back to his bottle and his fingers slowly whitened around its body.

That was his mum's default answer for the move, but it felt like they were running away. His dad's study was emptied into boxes and most of it never reappeared, as though the removal men were finishing the tumour's job and removing him completely. The few things that did make it into the flat didn't last long; the clockwork solar system in the living room being the last to go. One day it was standing on the windowsill, with the planets all lined up so that they would just fit on but could never turn, the next it joined Newton's Cradle and Neil Armstrong in their gloomy storeroom in the basement, supposedly to protect it from Alice.

The only remnants of his dad left on display were a few photographs, half-hidden behind open doors or candles, and only one of the whole family together. It stood in a silver frame on the mantelpiece and was taken around his bed not long before he died.

Danny sat on one side, his mum facing forward on the other so you couldn't see the bump, and the newborn Alice photo-shopped into his arms in between. Only his dad and Alice seemed to be smiling, like two bald and crinkly old men, though it was probably just wind. The doctors said Alice might come early as his mum was in her forties, but he still missed meeting her by a month. His mum said it was like she'd replaced him and he was still there in some way, but it wasn't … She didn't think it was either.

The first time he heard her was before the move, and he'd come down to the lounge door and put his eye to the gap between the hinges. She was curled on the sofa with her head in one hand, a drink splashing around the glass in the other, and her body shaking and snot dribbling from her nose. Alice started to cry, and as his mum started sniffing and wiping her eyes, he ran silently back to his bed. It didn't happen so often anymore but he still pulled the duvet over his head.

'You should get your father's science toys out again,' said Bert, clearing the dishes to one side. 'It will help you to remember, or better still, understand.'

'Maybe I don't want to remember,' barked Danny.

'Exactly. If you really understand something it becomes a part of you. And taking a fresh look at those toys will help you to understand the physics so that you *know* how to pass your exams without having to remember anything.'

'Oh,' said Danny, rolling back his head. He let his boots slide through the sand to straighten his legs, and blood rushed back to his fingers as the cola bottle fell against his thumbs. 'Yeah … maybe.'

'And just think,' said Bert, opening his briefcase onto the table. 'You could be the first to introduce the wonders of the universe to Alice.'

Danny looked up at the universe and tried to remember the first time he'd played with his dad's toys then realised they'd always been there. The brightest stars were already out, grouped in their constellations but with more appearing every second to hide them, and he wondered if he could find the Plough then leapt out of his chair.

On the Beetle's roof rack, to the whirring sound of a motor, the top of the white metal chest opened into two flaps, and a fat cylinder rose out and angled itself into the night sky, like a cannon.

'Wow!'

'My telescope,' grinned Bert. 'It is controlled by the computer, so all we have to do is type-in co-ordinates and the images are digitally recorded and displayed here on the screen. Have you used a telescope before?'

'No. Never.'

'Then you should climb up and look through the eyepiece first,' said Bert, his fingers tap-dancing on the keypad.

Danny climbed onto the top rail of the roof rack, and after motors had adjusted the telescope to point into the sky above the main Canyon wall, he put his eye to the small tube at its side.

Jupiter sat, a perfect white sphere in the blackness of space, her pale-orange stripes just visible together with the tiniest red dot. Four little moons sat around her; three white discs in space and one black shadow projected against her body, they appeared motionless in their orbits as though frozen in a photograph. But this was no captured image or flat projection, Danny could see that these were real bodies in the depths of real space, and he could almost feel the incredible distance between them.

'That is the view that greeted Galileo in 1609,' said Bert. 'The first man to look into the night sky with a telescope and see moons orbiting another planet, it set him on the path to understanding relative motion. Newton built on his work to formulate his three laws of motion and law of gravitation.'

Danny looked down from the telescope. 'What was Newton's Mount Everest?'

'Light,' said Bert, his eyes seeming to sparkle with a source of their own. 'The nature and behaviour of light and the strange mystery of its speed.'

'I know what it is,' said Danny, returning his attention to Jupiter and her moons. 'Nothing can travel faster than the speed of light.'

'That is merely one of the consequences. The real mystery of the speed of light is the fact that it has one. You see, in Galileo and Newton's time light was thought to be instantaneous.'

'What is the speed of light?'

'Incredibly fast. But it is still a speed, and if light takes time to travel distance, then that will affect our measurement of time.'

'Time?'

'Just as you do, Newton believed in absolute time. In his universe a body can move at any speed in any direction through space, but it marches forward through time to the steady beat of a great cosmic clock, ticking away at the same constant rate everywhere and with everything connected by a universal now.'

Danny pulled his head away from the eyepiece to find Bert peering up at him over the top of his glasses; his eyebrows rising like storm clouds. 'You mean that's wrong?'

'*There is no such thing as now.*'

Danny laughed. 'What about … NOW?'

'We each experience the passage of time as one now after another, but your now is not the same as mine or anyone else's.'

'Why not?'

'Because of the speed of light.'

Bert dropped a finger onto the briefcase keypad, and Danny nearly fell off the roof rack as a powerful lamp started rising telescopically beside him, drowning their camp in a sea of white light.

'Come on,' said Bert, springing out of his chair. 'Jump in and I will show you.'

Danny climbed down and reversed his bottom into his seat, as the metal pole swayed to a stop ten feet above his head. 'What's with the floodlight?'

'What can you see when it is completely dark?'

The black screens rose swiftly up the Beetle's windows, but no twinkling star field appeared and only Danny's voice emerged from the dark. 'Nothing.'

'We need light to be able to observe anything, but for centuries the eyes were thought to act like torches, somehow producing the light that enables us to see. It took the greatest scientist of the Middle Ages, an Iraqi known as Alhazen, to show that vision is light entering the eyes from the outside world. He did it using a camera obscura.'

A cone of light suddenly appeared from a tiny hole in Danny's passenger window screen, its shape clearly defined by the thousands of dust particles floating inside.

'Camera obscura is Latin for a 'darkened room', and if light enters a dark room through a small hole …'

Bert raised a rectangular piece of white card into the middle of the Beetle, catching the light cone in a blurred circle the size of a football. As he moved the card towards the window, Danny watched the random splashes of colour focus into an upside-down moving image of their camp.

He watched flames perform a hanging dance from the upside-down fire and embers fall from their tips as a breeze drifted by, gently swaying the top of the juniper tree at the bottom of the picture and ruffling the tufts of desert grass suspended along the top. It was clear in every detail, but there was no fancy equipment or lens producing this film; it was just a piece of cardboard and a hole in a screen.

'Whether a body emits its own light, like the fire,' said Bert, 'or reflects light, like the juniper tree, at each instant the light forms an image of the body which travels through the hole to appear on this cardboard screen. Replace the card with light-sensitive film and we could capture the light in a photograph. And just as photographs can be run together to produce a film, so the continuous stream of light reflecting off everything around us produces the continuous stream of events we see as daily life.'

'Why's everything upside down?'

'Alhazen realised it is because light travels in straight lines. The light coming from the bottom of the juniper tree has to travel on an upward diagonal to get through the hole and onto the screen.' Bert traced the diagonal path from the hole, visible as the top of the light cone, and his finger landed on the top of the card and the bottom of the tree.

'Light from the top of the tree travels in a downward diagonal,' said Danny, following the bottom edge of the light cone onto the bottom of the card and the top of the tree.

'And light travels in straight lines from every point in between to form a complete inverted image of the tree,' finished Bert.

'What's it got to do with there being no universal now?'

'Because if light enables us to observe the world, then its speed will affect *when* we observe it.'

'How fast is it?'

'So fast that when Galileo and an assistant tried opening and closing shuttered lanterns across a three-mile valley their hopeless results seemed only to confirm its instantaneous passage. But speed is simply the distance travelled per unit of time, so what was needed was either an extremely accurate clock, which of course they did not have, or to measure light travelling over a much greater distance.'

The light cone disappeared and a new source, the computer screen, bathed them in an orange glow.

'From Jupiter?' said Danny.

The planet was now five times the size, as though they'd rocketed towards her in Spaceship Beetle, her orange cloud belts imperceptibly rotating either side of the Great Red Spot and a single white moon just visible beside her.

'From Io,' said Bert, pointing to the moon. 'In the summer of 1676, a Dane called Olaf Roemer timed Io's orbit around Jupiter and found it to be 42½ hours. When he measured it again in the middle of winter it was over 16½ minutes longer.'

'Shouldn't it always be the same?'

'Yes, but Roemer realised that the different timings could be explained, and still accord with Newton's theory, if we accept that it takes time for light to travel across space to our telescopes. This would make our measurement of Io's orbital period depend on the Earth's position in its orbit around the sun and its distance from Jupiter. Using the fact that Io's winter light took an extra 16½ minutes to travel the diameter of the Earth's orbit, Roemer calculated the speed of light to be 140,000 miles per second. However, his measurement of the diameter was incorrect and we now know that light travels 186,000 miles every second.'

Danny stared at the Beetle's black windscreen, trying and failing to imagine that speed.

'Roemer showed that light travels at a finite speed and your timing of events, or *nows*, will not be the same as an observer's located at a different distance.' Bert nodded Danny's attention back to the computer screen. 'Keep your eyes on Jupiter. It is getting close.'

'What's getting close?'

'Using Newton's law of gravitation, astronomers have predicted that a large asteroid will collide with Jupiter at 8:15 pm on Monday, July 25th.'

'That's today.'

'That's in about 40 seconds. The asteroid has broken apart in Jupiter's strong gravitational field, but three sections will be coming in from the top-right of the screen at a relative speed of 134,000 mph any ... second ...'

The time code at the bottom of the screen counted away the last few seconds to 8:15 and three streaks of light burst across the screen, slamming into the giant planet and erupting into exploding flashes as they blew apart in the thick clouds of her upper atmosphere.

'Here on Earth, we have recorded 8:15 as the now of impact,' said Bert, as shock waves rippled out from the impact sites. 'Is that when the collision occurred?'

'No, it must have taken time for the light to reach us.'

'Jupiter is currently 390 million miles away, and it has taken the light from the impact, moving at 186,000 miles per second, 35 minutes to travel that distance and form the image for us here on Earth.'

'So it actually happened at 7:40.'

'To an observer on Jupiter it did, but what about an observer on the sun? The Earth is positioned between them, which means the light from the impact has a further 93 million miles to travel and it will not reach the sun for another 8.3 minutes, at 8:23. To reach our nearest neighbour, the star Proxima Centauri, the light will have to travel nearly 25 million, million miles and it will be four years before it is observed there.'

'But it has happened really, hasn't it? It happened on Jupiter at 7:40 this evening it's just they can't see it yet.'

'What does that mean? If the sun disappeared, right now, what would we notice?'

'Nothing.'

'That's right.'

'Because it's night-time.'

Bert's slow nod of the head changed direction, mid-fall, and became a few fast shakes. 'Even if it were the middle of the day, there would be no measurable effect because the Earth would continue to receive the sun's heat and light for the eight minutes it takes its final rays to reach us.'

'Only because we don't know.'

'We can only go on what we can measure and for over eight minutes there would be no measurable effect. The sun would still exist.'

'Yeah, but it has still happened.'

'Look at it this way. Because it takes time for light to travel distance, we are always looking backwards in time.'

'You mean we're seeing Jupiter as she was 35 minutes ago?'

'We cannot observe her as she is *now* because that is our now and we are seeing her as she was *then*. We will have to wait 35 minutes to see what is happening there now, and the further out we look, the further we look back in time. It takes light from the Virgo Cluster of galaxies over 60 million years to reach us, so the image of you, sitting here in my Beetle, has only just left on its journey and will not become their now for another 60 million years. Conversely, a Virgoan viewing us from their now would be more likely to see a Tyrannosaurus Rex.'

'There's still a now though, isn't there?' smiled Danny. It's just that we can't see somebody else's at the same time as ours.'

'Exactly. Now is a relative measurement that depends on your point of view. Even our nows are different. If we watched that juniper tree fall over, travelling 186,000 miles or a billion feet every second, its reflected light will take two billionths of a second longer to reach me because I am sitting two feet further away. You would see it happen before me and we would allocate different times to our different nows of it hitting the ground.'

'Isn't it like the time difference between London and Phoenix, and with the right adjustment we can figure out the correct time? The impact on Jupiter happened at 7:40, and if the sun did disappear we know it actually happened eight minutes ago.'

'What do you mean by correct time? Whose time? In theory, you could be right and time is essentially passing at the same rate everywhere as a series of universal nows, except for one thing. There is no such thing as standing still.'

'What difference does that make?'

'We have been talking about the distances between bodies as though they are constant, but everything is moving. That brings us to the real mystery of the speed of light: the fact that it has one.'

'We've done that.'

'I mean, the fact that it has *just* one. All movement is relative, so how can anything have only one speed?'

Danny opened his mouth then closed it again, staring at Jupiter on the screen as Bert drew a long torch from the pocket of his shorts, like Mary Poppins producing a hatstand from her carpetbag.

'Sitting there now,' said Bert, shining the torch past Danny's head, 'you will measure the speed of this beam to be approaching you at 186,000 mps. If I whiz around the Earth in Spaceship Beetle at half the speed off light and come hurtling back towards you, at what speed will you measure the torch beam to be approaching?'

'One-and-a-half times the speed of light.'

'Newton's addition of velocities says it should be 279,000 mps.'

'But nothing can travel faster than light.'

'Which is why you will still measure it to be 186,000 mps. If I turn the beam back towards you as I rocket past, even though Spaceship Beetle is moving away from you at half the speed of light and Newton's subtraction of velocities says you should measure its speed to be only 93,000 mps, you will still measure the beam to be approaching you at the speed of light. Whenever we measure the speed of light the answer is always 186,000 mps, regardless of how the source of that light is moving.'

'That's weird.'

'Just how weird depends on what light is made of and how it travels through space. Based on his laws of motion, Newton published the first full theory in 1704 saying that light shoots out from a source as a stream of tiny particles, like bullets from a machine gun. This explains shadow formation and how it bounces off surfaces like balls bounce off walls. But in 1800 an Englishman called Thomas Young performed a simple experiment that showed light behaving as a completely different animal.'

Bert tapped a key and the computer screen blinked off, returning the Beetle to darkness. 'If you fire thousands of machine gun bullets at a slit in a wall, where would you expect them to hit a target on the other side?'

'They'd be grouped opposite the slit. In a slit-shape.'

'So if you pass a beam of light through a tiny slit,' said Bert, raising the cardboard screen.

A projection of light appeared from Danny's window screen and formed a tall, thin rectangle on the card.

'Where's the camp?'

'I have redirected the floodlight onto the car so that we are looking at a direct source. As Newton's theory predicts, that produces a rectangle of light on our card. What will two separate slits produce?'

'Two rectangles of light.'

A second projection appeared, forming a second rectangle next to the first.

'Now if we bring the slits closer together so that the rectangles of light overlap, what will that produce according to Newton?'

'A larger, brighter rectangle?'

Danny was expecting something unexpected, but he didn't expect to see a series of vertical bands appear on the card, alternating between darkness and light, like an old car radiator grille.

'Have you ever seen bullets do that?' said Bert. 'The combined light from the two slits has organised itself into separate stripes of light. Young's simple experiment shows that light does not move as a stream of particles, light moves as a *wave*.'

'Yeah, a light wave,' said Danny, as the window screens lowered to reveal their camp, bathed once again in only fire and starlight. 'How does that help?'

'There is only one way to find out,' said Bert, opening his door. 'Let's make a wave.'

Raising his arms, jumping out of his seat and singing 'Wey-y' as he bounced off the roof, Danny flopped everything down to discover he was alone in the Beetle.

10

It was easy to lose track of time inside Spaceship Beetle and Danny had forgotten there was a warm, fire-lit night waiting beyond its screens. He climbed out and lowered his bottom into his chair and his arm into the cool-box next to it.

'Energy is not destroyed in a collision, it is passed on as a moving wave,' said Bert, unfurling a thick coil of luminous-red rope along the ground. 'But for the energy to move there has to be some form of connected substance, or medium, supporting the wave.'

Bert raised his arm up and down, and Danny watched the bulge of a wave travel the length of the rope, like a fiery-red serpent rearing its back and whipping its head into a lunging attack.

'That was one rope wave,' said Bert, creating another. 'It looks as though the rope itself is moving forward, but only the kinetic energy of the up and down vibration is transmitted as each part of the rope passes it on to the next part, pushing it along to form the crest and trough of a moving wave.'

'Like a wave at a football stadium. The people move up and down but the wave moves forward.'

'Let's look at a few of them.'

Bert tied the end of the rope to the juniper tree and started waving his arm up and down at a steady pace, resurrecting the serpent in its classic shape: a wavy line pulsating luminously above the ground.

'If the input of energy is constant, a series of crests and troughs line up along the length of the rope forming a continuous wave made-up of identically-sized vibrations or waves. In an earthquake a vibration in the Earth's crust is transmitted through the ground as a seismic wave. When wind lifts an area of water into a crest and gravity causes it to fall again, this alternating kinetic and gravitational energy is transmitted as a water wave. And we are able to have this conversation because every time you speak you deliver controlled bursts of energy that vibrate the air in precise ways to produce the alternating high- and low-density pressure waves travelling out of your mouth as sound. But no matter what medium it is travelling through, all waves move as a series of crests and troughs and all waves exhibit a property called *interference*.'

'Like you get on the radio?'

'That's right,' said Bert, dropping his arm to his side. 'Waves can interfere with each other in different ways, depending on how they meet. If identical waves meet such that their crests and troughs line up then they are said to be in-phase, which means they add together or reinforce each other. This *constructive interference* intensifies the energy of the wave.'

That made sense to Danny because the waves were combining together.

'If identical waves meet such that the crests of one line up with the troughs of the other then they are said to be out-of-phase. This *destructive interference* causes both waves to disappear altogether.'

That didn't.

'It helps to think in terms of equal-sized positive and negative energies cancelling each other's effect. If you chop the crests off water waves, turn them upside down and drop them into the troughs, you would be left with flat water. With sound waves, if the high-density crests of one pressure wave line up with the low-density troughs of an identical second pressure wave, then the two add up to normal pressure again and there is nothing to hear.'

'Is that what happened with the light in Young's experiment?'

'Exactly. The bands of light on the card corresponded to areas where the light waves reinforced each other due to constructive interference, and the dark bands where they cancelled each other out due to destructive interference and there was no light. According to Young's experiment, light travels as a wave.'

'So Newton was wrong?'

'Light behaves like a wave, but then it often behaves like bullets. They are both just theories, and if he was to topple Newton, Young needed more evidence that his was right. It arrived in 1864 when a

Scottish mathematical genius, James Clerk Maxwell, discovered four equations that predicted the existence of electromagnetic waves and the constant speed at which they would travel.'

'186,000 mps?'

'Now the logical power of mathematics was confirming that, just like sound, light is a form of wave with a fixed speed.'

'Hold the horses. Why's sound only got one speed?'

'All waves move at a set speed that depends only on the medium through which they travel. It varies a little with temperature and pressure, but sound waves move through air at 760 mph. Water waves move through fresh water at a set speed determined by the depth of the water. And rope waves move at a set speed determined by the molecules and construction of the rope. We can see why, if we measure the energy of the wave.' Bert started waving his arm up and down. 'Count the number of wave crests that pass you each second.'

Danny counted the long waves travelling past him down the rope. 'It's about one crest every second.'

'My arm is going up and down once every second; that energy creates a new vibration or wave every second; and those waves travel along the rope and pass you at a rate or *frequency* of one wave per second. If I put in more energy and jig the rope up and down twice as fast?' Bert doubled his arm speed.

'It's now two crests every second.'

'I have doubled the input of energy and that has doubled the frequency to two waves per second.' Bert dropped his arm to his side. 'The greater the energy, the more waves are created and moving each second, and so their frequency is a measure of that energy.'

'But why is the speed of each type of wave always the same?'

'Speed is simply the distance travelled per unit of time, and another feature we can measure for any wave is its length. Since we count waves from one crest to the next, we say the *wavelength* is the distance from one crest to the next. What do you notice about the wavelength when the frequency of the wave changes?'

Bert started waving his arm at one wave per second and a few, long waves travelled along the rope. He raised the frequency to two waves per second and more, shorter waves travelled, squashed together, down the rope.

'The wavelength goes down,' said Danny.

'And by a proportional amount. If you double the frequency or number of wave crests per second –'

'You halve the wavelength of each wave.'

'Which means, regardless of their frequency, the waves always travel down the rope at the same speed. For our rope wave, if one wave produced at a frequency of one wave per second has a wavelength of, say, eight feet then it will travel at eight feet per second. Two waves produced at a frequency of two waves per second each have a wavelength of four feet, which means –'

'Two four-foot waves still travel eight feet in one second.'

'And if we double the frequency again to produce four waves per second,' said Bert, forcing his arm up and down as fast as he could.

'Four two-foot waves pass me every second.'

Bert dropped the rope and tottered back to his chair. 'It is the properties of the medium through which it travels that determine the speed of a wave, not the input of energy. Whatever its frequency, the same distance will be covered in the same interval of time and the speed of all waves through the medium is the same. For this particular rope it is eight feet per second.'

'So a bang travels at the same speed as a whisper.'

'Ha! That's right.'

'What if the sound's moving, like when that lorry shot past blowing its horn?'

'It does not matter how the source of a sound wave is moving relative to you, its speed is governed by the medium of the air. If I run towards you clicking my fingers regularly, each click creates a sound wave that has less distance to travel to reach you. But each still moves at 760 mph from the place it was created and set in motion, which is why you hear each click a little sooner than the last. The same applies if I am travelling away from you, except that the sound wave of each successive click will take longer to reach you at that set speed of 760 mph.'

Danny bobbed forward on his chair. 'That explains it then. Light is a wave and it travels at a fixed speed of 186,000 mps relative to …'

'Exactly!' cried Bert. 'Relative to what? If light is a wave then what is waving? What is the medium transmitting light waves across the vast distances of our universe at 186,000 mps?'

'Space?'

'But space is not something, space is nothing. I will show you.'

Bert lifted the wooden BJ crate onto the table, slowly raising the entire top to reveal a tall glass dome sitting on a polished mahogany base. It looked to Danny like the housing for an ornate mantelpiece clock, but a grinning Mickey Mouse alarm sat small at its bottom, with vertically mounted bells for ears and his eyes tick-tocking from side-to-side.

'This is the bell jar experiment, first performed in the early 1800's,' said Bert, placing the glass dome on the table, cushioned by the rubber seal running around its rim. 'Using this tube, which attaches to a hole in the bottom of the base, and this simple pump, we can suck out all the air and create a little piece of space here on Earth.'

Raising a long, orange tube in one hand, a two-handed stirrup pump in the other, and an extra big grin in between, Bert started sticking them all together. 'Right,' he said, handing Danny the pump. 'When the alarm rings start pumping as fast as you can so that we remove all the air before it stops.'

'Okay,' said Danny, standing up and wiping his hands down the sides of his jeans.

Bert fiddled with the back of Mickey and placed the bell jar back on its base. And with his feet on the stirrups at the bottom of the pump, his hands on the twin handles at the top, his back bent down, his elbows up, and the plunger pulled to its full extent ready for the first descent, Danny waited.

He waited … and waited … and waited …

'When did you set it for?' he said, standing back up at the exact moment that Mickey's bells went off.

Being surprisingly shrill for such a small clock, Danny leapt off the stirrups at the same time as he pushed down with the one hand still holding the plunger.

'Hurry!' said Bert, as Danny scrambled in a heap at his feet.

Danny caught the plunger handle behind one of the table legs, in his panic to get it out from underneath, and after Bert had grabbed and steadied the bell jar, he got back into position and started to pump.

Concentrating on a furious early rhythm, it wasn't until he was straining through the last inches of the final push that he realised the bells had stopped ringing. It wasn't until he looked at the alarm clock, then bent down and peered at it closely, that he realised they hadn't. Not a breath of sound was coming from the jar, but Mickey was vibrating around the base, the hammer a blur between his ears.

'Do you see?' said Bert. 'We have created a vacuum inside the jar and there is nothing to transmit the vibration of the bells.'

'So we can't hear anything,' said Danny, still smiling at the fact that they couldn't.

'Space is the same. It is a frictionless vacuum, which is why the planets and their moons remain in continuous orbit, and yet look. We cannot hear them but we can *see* the hammer hitting the bells, so what are the light waves travelling through at their fixed speed of 186,000 mps in the vacuum of the bell jar or empty space?'

Danny watched Mickey silently vibrate to a halt and shrugged. 'There must be something still in there.'

'That must be the answer. So scientists put forward the idea that the vacuum of space is filled with a substance they called Luminiferous Ether. Just as clicking my fingers produces vibrating sound waves in the air, so lighting a lamp was believed to produce vibrations in the Ether.'

'What's Ether?'

'Nobody knew. It was supposed to be this invisible, frictionless substance that is fixed and motionless everywhere in the universe and through which everything moves without noticing.'

'Then how do we know it's there?'

'The clue lay in the way all waves behave. When I run towards you clicking my fingers –'

'Each sound wave travels to me at 760 mph from the place it was created.'

'So motion of the source through the air does not affect the speed of a sound wave. But what if a wind is blowing that air towards you at 10 mph? The speed of the sound wave relative to the air will still be 760 mph, but the speed of approach of the sound wave relative to you on the ground will be the addition of those velocities.'

'770 mph.'

'What if the wind is blowing away from you?'

'You subtract the velocities and it'll be approaching at 750 mph.'

'What if instead of the air moving, *you* are moving through still air at 10 mph towards the source of the sound?'

'Like sticking your head out of a car window, it would be the same as standing still in a 10 mph wind.'

'So the approach speed of a sound wave will depend on how you are moving because your movement is identical to creating the effect of a wind. Now we can see a way of detecting the Ether. If the speed of light is always 186,000 mps relative to the motionless Ether, then an observer who is at rest relative to the Ether will always measure that to be its speed, regardless of how the source of the light is moving.'

'Oh, but observers moving relative to the Ether will see light waves coming towards them at different speeds, depending on how they're moving.'

'In 1887 two Americans, Albert Michelson and Edward Morley, split a beam of light into two perfectly aligned light waves and sent them on equal-length journeys; one travelling in the direction of the Earth's orbit, the other at right angles. Because the Earth must be moving though the Ether as it orbits the sun, the light beam moving in the same direction should be heading into an Ether wind and travel more slowly than the right-angled beam. Having started perfectly in-phase as one beam, if they were out-of-phase after being recombined on their return, both having covered the identical distance, then they must have travelled at different speeds due to the effect of the Ether wind.'

'What happened?'

'Nothing.'

'There was no interference?'

'Not a hint. They repeated the experiment over subsequent months to counter any possible effect due to the rotation of the solar system, but the speed of the light waves was always the same, regardless of the direction they travelled. That could only happen if the Earth was stationary in the Ether when we know it orbits the sun.'

'What does that mean? There's no Ether?'

'It means we have uncovered the real mystery of the speed of light. If Morris' Citroen behaved like a wave of light moving at 30 mph, then driving towards it in my Beetle, regardless of our speed, we will measure it to be approaching at 30 mph. If we come to a stop and measure its approach speed again, we will still measure it to be 30 mph. When it goes past, if we chase after it at five mph, 29 mph or any speed in between, we will still measure it to be approaching at 30 mph.' Bert held up his forefinger, arrow straight in the air. 'There is only one measurable speed of light and light travels at that one absolute speed *relative to everything*.'

'What about the addition and subtraction of velocities?'

'A beam of light does not follow the addition and subtraction of velocities. Or the velocities always add and subtract to produce the same figure, which is not how relative motion works under Newton at all.'

'That is weird.'

'The fact remains, as Maxwell's equations predict, the speed of light is absolute and does not conform to our understanding of relative motion.'

Bert emptied his bottle of Apache and started clearing the table. 'That is the mystery of the speed of light; that is the conundrum that led Einstein to his theory of relativity; and that is where our expedition heads tomorrow. Grab the sleeping bags from the Beetle. I think they are under your seat.'

'Did Einstein find the answer?'

'Yes. His theory of relativity explains why light waves do not need a medium like the Ether in order to travel; why they always move at their one fixed speed relative to all observers; and why nothing can travel faster than the speed of light.'

As far as Danny was concerned, that was still the weirdest aspect of all. 'Bert, why can nothing travel faster than the speed of light?'

*

Danny sat on his sleeping bag and started to remove his boots, until it occurred to him that, if nothing else, at least he could sleep like a cowboy. Crossing his boots at the ankles, he lay down beside the glowing embers of the fire, but all thought of missing hats vanished as he linked his hands behind his head.

A zillion stars shone down from between curtains of black canyon wall, each transmitting its history down a time-beam to his eyes. Fading in brightness to unfathomable depths, hinting at the unbelievable size of it all, they clustered in star fields, each twinkle a tiny individual sun, and merged into galaxies and nebulous wisps of light. And rising across the entire night sky, a towering grey cloud, explosions of light shining through and around its edges, like smoke from an industrial chimney drifting in front of the sun.

'Isn't it amazing?' said Bert, his outline frosted with starlight. 'That stripe of lighter sky is a flat disc of billions of stars and our home galaxy, the Milky Way. As we look towards its centre from our position on the outer edge of one of its spiral arms, we see the faint light from each star build up, one behind the other, combining and concentrating to create a band of light.'

Bert settled into silence then leapt back out. 'There are more stars out there in the known universe than there are grains of sand on all the beaches of the Earth. Imagine that! And to think, from this little planet, using only mathematics, experimentation and the power of our minds, we know how the universe evolved from almost the instant of its birth.'

'The Big Bang. Is that what got everything moving in the first place and gave it inertia?'

'That's right. Thirteen point eight billion years ago a primordial atom of pure energy exploded, creating matter, space and time and pushing them outwards, like the surface of an inflating balloon, to form the billions of galaxies and vastness of space still expanding in all directions to this day.'

'There's Jupiter,' said Danny, pointing to the brightest object in the sky.

'It is about to pass Vega, the second brightest star in the northern celestial hemisphere.'

'How far away is Vega?'

'If you could travel at the speed of light, it would take you 25 years to reach her.'

'Fifty years for the round trip. That's nearly a lifetime.'

'Yes,' said Bert, softly. 'Fifty years is a long time.' He lowered himself back to his sleeping bag. 'But is it a lifetime gained or a lifetime lost.'

Danny imagined he was exploring the universe in a spaceship, zooming towards Vega with the stars streaking by. 'We need to invent a warp drive, like on *Star Trek*,' he said, after what seemed like only a minute. 'Do you think we ever will?' He looked over at Bert lying still in his sleeping bag, like an Egyptian sarcophagus awaiting burial in the sand. 'Bert?'

Twenty minutes later – after he'd counted five satellites and three shooting stars in a game of galactic *Where's Wally* – Bert suddenly broke the silence.

'Danny.'

'You're awake.'

'I know that my expedition is not what you expected of your science camp.'

'Eh? Well … no. But –'

'In a way, it is good. It has given me the chance to think again. Tomorrow morning I will drop you off at the Missing Horse and you can spend the rest of the week there.'

'What do you mean, drop me off?'

'Morris will show you how everything you have learned with me directly translates to your physics syllabus at school. He runs regular workshops for Coconino High so there will be other students your age and plenty to do. The horses need exercising daily but the ranch hands will get you up to speed, and I'm sure your colleagues will show you the local haunts.'

'Where will you be?'

'I am stopping for fresh supplies but I will not be staying … I think it is best I continue Stage Two of my expedition alone.'

'Oh.' Danny watched a shooting star blaze across the sky, like a space shuttle in fiery re-entry, but he didn't add it to the count. 'I thought you needed my help?"

'I do. More than in Stage One. You have been the perfect companion, helping me to critically analyse the base concepts that underpin all of cosmology. But I am now thinking that the whole idea is the confused desperation of a tired old man and taking you with me would be a selfish act?'

'Why?'

'In case something happened to you.'

'Like what?'

'The same thing that happened to me.'

'Bert, what did happen to you?'

'No. I have decided it is best you do not get involved. Ha! And you do not want to end up looking like this.'

Danny pictured Bert's grinning face onto the silhouette of his head. 'What about the mystery of the speed of light?'

'I have no doubt, one day, you will find the answers.'

'Yeah, like that's going to happen.'

'I am not taking you with me because I think it is too difficult for you to understand. You are more capable than most adults, weighed down as they are with the prejudice of a lifetime's misexperience. I am not saying it will be easy, either; abandoning what you believe to be true never is. But the answers are out there if you want to find them.

'What if I want to find them with you?'

Bert sat up, like a reincarnated Mummy. 'Oh! Well that would be different ... if you really want to come.'

'What would we do, just drive around more? And what might happen –'

'You do not need to make a decision now. Let's get to the Missing Horse and you can meet everyone. When the Beetle is resupplied, we are rested and refreshed, and I am holding open the passenger door, only then will I ask you: fifty years from now, when you're looking back at your life, don't you want to be able to say you had the guts to get in the car?'

'Isn't that from the first *Transformers* movie?'

'You do not realise but you are still wearing a blindfold – one that has been wrapped thickly around your senses since the day you were born. But now, armed with your new understanding of inertia, gravity, energy and wave motion, you can follow Einstein's path. You can climb into our spaceship and travel 13.7 billion years back to the Big Bang and the beginning of time. There, Einstein will remove the blindfold and reveal an amazing new world to your eyes. A world that will challenge your understanding of space and time … a world where your reality is nothing like it appears."

Bert settled back down to the sand and the stars. "The choice is yours."

Einstein's Beetle

Part II

Einstein's special and general theories of relativity, $E = mc^2$ and the Big Bang

11

Danny spent the night in complete mental shutdown. As a result, the last thing on his mind before drifting off was the first thing to occur to him on waking and he felt like he hadn't slept in between for thinking about it.

The shadow of the Canyon's eastern wall had just cleared camp and he stretched in the morning sunlight, like a reptile warming its blood.

'Bert?' he called, scratching at various bits as he wandered over to the table.

He scanned the canyon floor and its horizon with the limestone walls, dropped onto a chair, and dived for the note taped to the Beetle's passenger window.

> Off doing an experiment.
> I have fixed the Beetle
> so we can leave as soon
> as I get back, around
> mid-morning. Hope you
> do not mind. Help your-
> self to anything you
> want.
> Bert

Bert had written the message over a simple map showing you how to get somewhere, and the somewhere, Danny realised, was Joe's mum's motel, Rosie's. He peeled the brochure off the window and noticed the key sitting in the Beetle's ignition, and by the time he'd read the description of her Full-English breakfast, he knew exactly what he wanted and it wasn't in the cool-box.

He re-taped the brochure to the table and scribbled on its other side:

Off doing an experiment
with my stomach at
Rosie's - see how much
it can hold. Hope you
don't mind. You did say
anything!
      Danny

Five minutes later, he was sitting behind the Beetle's steering wheel sucking on a cherry-cola lollipop and turning the key in the ignition.

The computer beeped into life, but nothing happened – which was good because nothing had ever seemed to before. He pressed the accelerator pedal; the Beetle started moving; and they bounced across the old riverbed onto the narrow dirt track, skirting the scree slopes of the south canyon wall as they followed its path down to Rosie's.

Danny grinned. Everything was going to plan. He'd drive as far as the highway; it was only a short walk from there. If he saw anyone he'd stop and scoot over to the passenger seat, and if he ran into Bert he'd say he'd come to see if needed any help.

Unfortunately, Danny's plan hinged on him being in control of the Beetle, and deciding to show that he wasn't, she raised her black screens up the windows, beeped '**0 mph**' onto the computer screen and dropped her brake pedal, with a clang, to the floor.

'Oh Ma-an. Not again,' he wailed, clicking home his seat belt and tightening the strap.

The roof-light blinked on and he spent two minutes wincing under its pale yellow light, with one foot braced against the dashboard and his hands around the frame of his seat. Lowering his leg, he spent the next ten trying to get out of the Beetle, with his initial pokes at the keypad and commands '*End spaceship-mode*' and '*Lower window screens*' progressing to a volley of fingers firing at any and every key whilst shouting '*Let me out of here, you psychotic heap of scrap!*'

The computer beeped, halting his attack, and as '**Time 09:00**' and '**Speed 161,000 mps**' materialised on its screen, a brilliant-white spotlight appeared in the driver's window, streaking beams in all directions, like the star of Bethlehem.

Danny shielded his eyes and blinked at his reflection in the windscreen. Suddenly unclipping his seatbelt, he twisted round and the stick of his lollipop fell against his chin.

A semi-circle of wispy white clouds floating above the green mass of the America's, with the blue Atlantic disappearing over the curving horizon on one side and a sharp shadow line off the western seaboard plunging the Pacific into darkness on the other, planet Earth sat in the rear window, huge, beautiful and shrinking fast.

Danny smiled and squinted back at the giant star of the sun, watching it crawl across the windows as the half-lit Earth continued to shrink in the rear. He didn't notice the star field in the windscreen until the Earth was a blue speck in the blackness of space and the Beetle's sloping roof had blotted out the sun, but he knew where they were heading as soon as he pressed his nose up to the glass.

The light from a million stars filed between his hands, not twinkling as before on the screens but as tiny circular suns, and as the Beetle zoomed silently through space, an orange dot crept towards the middle of the windscreen.

At the size of a pea, the red spot appeared the size of a pinprick. At the size of a beach ball, the orange and white cloud belts became powerful jet streams, circling the planet in opposite directions and whipping their boundaries into vast, swirling storms.

As Jupiter continued to grow, Danny marveled at the quality of the computer-animation. The colours seemed more natural than the enhanced images from the telescope, with a depth that made everything appear real. He could see reflections bouncing off the Beetle's bonnet, as though he were looking through the glass, and now and then his stomach would lurch with the sensation of motion.

The giant planet's outline disappeared off the edges of the

windscreen, the Great Red Spot expanding in its place, and just as Danny thought they couldn't possibly get closer, and all he could see was swirling red cloud within red cloud and raging storm within storm, the computer beeped, the clock hit '09:20' and they didn't.

Spaceship Beetle had reversed direction, and twisting back and forth in his seat, Danny's attention moved from Jupiter, shrinking fast in the windscreen, to the blue dot with its attendant star imperceptibly growing in the rear window – via the box of cheese and tomato bagels he found on the back seat.

The computer beeped and a voice started squeaking from the speakers, as though it had inhaled a balloon-full of helium, and Danny turned to find Bert's face on the screen, twitching as though connected to the grid.

The video appeared to be stuck in fast forward, but kept repeating in a loop, and picking out the words *key, trouble, hold down* and *red,* Danny deciphered the message and put his finger to the red key. He'd already pushed it countless times – it was the first one he tried – but this time he kept it pressed.

The computer beeped again; the Earth started visibly growing in the rear window; and the sun moved rapidly across to its left-hand edge, disappeared behind the roof strut and started crossing back along the side of the Beetle.

As the Earth reached the size of a marble, he could see the great mass of Asia and Europe rotating out of the hemisphere of shadow, with the deep blue Atlantic following right behind. As she suddenly expanded to fill the entire windscreen, the briefest impression of the Americas blinked through his brain, the windows turned black, the roof light yellow, the clock hit '09:40' and the speed beeped to zero.

The ride appeared to be over.

Danny tried his door, but with a new appreciation for spaceship mode he didn't mind that it remained locked. He wasn't surprised when five minutes later, and without any action on his part, the window screens suddenly lowered, but he was by the sunlight that came flooding over the top of them when they did.

He fumbled his way out of the Beetle and leant against the roof rack, rubbing his eyes until they focused on its cargo of crates. Realising that he wasn't standing on dirt, and reasoning that his next move should be to check, he refocused onto his feet and the uneven surface of white limestone that appeared to extend underneath the Beetle. Pushing off the roof rack, he span round and his brain abdicated all responsibility for further reasoning.

The Beetle sat inside a crescent-shaped limestone wall, which enveloped her back and most of her sides as though she'd reversed into a crumbling double garage. The stone floor ran level for ten feet then dropped three to a vast plain with a distant range of hills rolling along its horizon, like a bank of fog. There was the usual mix of sand and spindly covering of scrub, but no dirt track, no scree slopes … and where were the towering Canyon walls!

Seconds later, he'd scrambled up the loose rock at the rear of the Beetle's garage and found them.

*

Long ago, when dinosaurs roamed the Earth, the seabed rose to form the Colorado plateau, 12,000 feet above sea level, and a mighty river started carving a canyon into its soft sedimentary rock. As tributaries gouged side-canyons into its collapsing walls, side-canyons bit into their collapsing walls, on and on with the river removing half-a-million tonnes of earth every day. Sixty-five million years later, when it was 270 miles long, up to 29 wide and with 25 distinct rock layers exposed to a depth of over a mile, Spanish conquistadors stumbled upon its edge and named it *El Gran Cañón*.

Gazing into the abyss, Danny didn't know any of that. The Colorado River was only visible here and there and looked like a stagnant trickle in a winding gully. But as the Canyon's far wall plunged into the earth, the same staggering distance on the river's other side, tabletop mountains soared from its floor; their white island surfaces level with his boots. Smaller mountains rose all around them as great temple mounts, giant-stepped pyramids and towering cathedral spires, isolated from each other by huge expanses of space but joined by the same bands of colour. Purples on top of greys, yellows on top of reds of every hue, and topped with a thin layer of white, they rose like the stacked plates of a giant 3-D puzzle and a monument to the inexorable passage of time.

Most visitors are captivated for days by their first view of the Grand Canyon, some for weeks and some never leave. Danny lasted three and a half minutes before springing to his feet. 'Jumping Geronimo! How did I get up here?'

Hearing a rumble behind him, he turned as 'Yee-ha!' arrived through the air, and two horses trailing a cloud of dust galloped towards him over the ground; a hunched cowboy whipping the white horse's reins across its flanks with his left hand and leading the packhorse beside him with his right.

The horses locked out their front legs and skidded to a stop at the base of the rock ledge holding the Beetle. And dressed all in black, like Wild Bill Hickok opening one of his shows, Bert reared his horse to wave hello with its front legs as he cried 'Danny!' and waved hello with his hat.

Danny laughed and scrambled along the garage wall; jumping the last few feet to the sand as Bert did the same onto him.

'Thank goodness,' said Bert, trapping Danny's arms to his sides in what felt like more of a wrestling move than a hug.

'I'm sorry I took the Beetle, Bert.'

'Do not worry about that. The important thing is you are back.'

'Yeah, how did I get up here?' said Danny, pointing towards the Canyon with his leg.

'Ah.' Bert's arms dropped to his sides and flapped once. 'First things first,' he said, jumping his forefinger into the air. 'You must be hungry.' He headed for the packhorse and the cool-box strapped to its back; the chairs, small table and assorted saddlebags hanging on either side.

'No, I found some bagels in the Beetle. Bert, how did I end up here?'

'My breakfast. I forgot it on the morning you left; that's why I came back early.'

'You mean this morning.'

'Grab this,' said Bert, loosening the leather straps from around the cool-box.

'I'd only been driving for a couple of minutes,' said Danny, taking the weight and lowering it to the ground. 'She suddenly went into spaceship-mode.'

'Oh and look!' said Bert, pulling one of the saddlebags to the ground and opening its flap.

'Then I did the flight to Jupiter, but when I got out the track and everything had gone and somehow I was up here when we were down there.'

'You will need this,' said Bert, pulling out a wide leather belt and throwing it to Danny. 'And this. Oh and this, if the wind gets up.'

'And I mean a long way down there,' said Danny, catching the tan leather vest and blue necktie as he directed Bert back to the Canyon with the belt. 'What are these for?'

A pair of leather chaps passed, 'Where did you get those horses?' in mid-air, but when a grey Stetson followed, the questions stopped. 'Wow! Thanks Bert.'

After Bert had adjusted everything from his chaps to his Stetson, Danny stood tall, his hands hanging loose at his sides. Spinning into a half-crouch, he drew two imaginary six-shooters and emptied both chambers. 'I've always wanted to be a cowboy,' he said, re-holstering his fingers.

'We get a lot of those round here.'

Danny lost his balance forward and swaggered after it, his arms bent at the elbows and cycling round with his legs.

'Now what are you doing?'

'I'm walking like John Wayne.' Danny sidled to a stop, but as he tilted back his Stetson with a flick of his finger, his body straightened up and his arm pointed down into the Canyon. 'Bert, what's going on?'

Bert sighed deeply and remained motionless for some time. 'You are not going to believe me,' he said at last, trying to lessen the sadness in his eyes with a smile and failing miserably. He trudged round to the white horse and reappeared a moment later with his briefcase. 'You have been on a journey,' he said, opening the case onto the limestone platform.

'Yeah, those 3-D screens are brilliant.'

'That was not the work of the screens, Danny. They can create a star field and simple images of Spaceship Beetle II but that is all. Watch and I will raise them.'

Jumping back up to join the Beetle, Danny watched black, plastic-like screens rise swiftly from wide slots between her body and the glass. A thicker, large rectangular section sat in the middle of the windscreen with a wide line running horizontally from it, and across both the driver's window screens, to a small circle on the rear screen.

'LED's are buried throughout the screens to simulate a star field,' said Bert, 'and the raised areas are a series of connected monitors. They displayed the computer animation of the growing Spaceship Beetle II and the image of her racing past and disappearing into the distance.'

'What about Jupiter? And the Earth and the sun?'

Bert shrugged. 'I can program simple images to play in those specific areas, but not a full screen, 360-degree view of a trip to Jupiter and back.'

'Then how did you do it?'

'I did not do anything. You have actually travelled that journey.'

Danny waited for Bert's jaw to drop into a grin, but the sadness in the old man's eyes seemed only to deepen.

'I only designed and programmed my Beetle to simulate relative spacetime, but it really happens.'

'What really happens?'

'All movement is relative, so if a body is moving relative to us, we know we are moving at the same speed in the opposite direction relative to it.'

'Yeah, so?'

'So the FSP, my Flight Simulation Program, finds a body that is moving in the opposite direction we want to go and my Beetle experiences that relative motion as though we were looking from that point of view.'

Danny laughed. 'Like that can happen.'

'Thank goodness the upgrade to my rotational-synchronisation software worked and you only ended up here.'

Danny stopped laughing. 'Bert, how did I get up here?'

'You have been on a journey, but you have not travelled through only three dimensions; you have travelled through four. You do understand our three dimensions?'

'I've got a pair of those glasses at home.'

Bert lifted a trouser leg and drew a 12-inch wooden ruler from his black leather cowboy boot. 'If we want to measure the location of a body then the number of dimensions will affect how we can do it. A one-dimensional world would be like an infinitely thin piece of string.' He bent down with the ruler and drew a single line in the sand between his feet. 'Beings living on such a world can move backwards or forwards in one direction, but they can never pass each other or play leapfrog because there is no left and right or up and down in which to move. To find the position of a body in that world you need only measure its distance from one end.' He drew an X on the line and then turned it into a square. 'What about beings living on a two-dimensional world, like an infinitely thin sheet of paper or the surface of the Earth?'

'They can go wherever they want.'

'But they cannot play leapfrog because there is no up and down,' said Bert, drawing a little stick figure inside the square. 'How many coordinates do we need to pin him down?'

'You measure the distance from two of the sides.'

'We need two measurements from any two adjacent sides.' Bert jammed the ruler into the sand at one corner of the square, turning it into an imaginary box. 'How many coordinates do we need to measure the location of a body in our three-dimensional world?'

'Three. The distance from two walls and its height above the ground or from the ceiling.'

'What about the position of a fly buzzing around the room, or an airplane in flight or the Moon in orbit?'

'But they're moving. Everything's moving.'

'Exactly. Everything is constantly moving through both space and time, which means the position of a body is not a place in three dimensions, it is an *event* in four dimensions.'

'We need to say when we make the measurement as well.'

'Time is the fourth dimension and it is interwoven with the three spatial dimensions to form one, combined *Spacetime Continuum*.'

'I've always wandered what that meant.'

'Your whole life experience has prejudiced you against this picture, but it means there exists a deep relationship between space and time and they are not the separate entities you think them to be.'

Danny wasn't sure what picture he was meant to be prejudiced against but he figured it was the same for everyone else. 'What's it got to do with me?'

'We are all constantly moving through four-dimensional spacetime,' said Bert, pulling the ruler from the sand and pointing it at Danny, 'but you alone are up here when you were down there. The relative position of everything did not magically change for me.' He nodded to the white horse. 'I got here on Marilyn.'

'How *did* I get up here?'

Bert placed his hands onto Danny's shoulders and looked him straight in the eyes, 'You have experienced that we live in a four-dimensional universe where the speed of your motion through space affects the speed of your motion through time.'

'What do you mean, my motion through time?'

Bert squeezed Danny's shoulders then gave him something completely new to worry about. 'The horses need walking down,' he said, stripping the last load off the packhorse and handing its reins to Danny. 'So come on, cowboy. Let's go for a ride.'

12

Stripped of his unflattering load, Charlie the packhorse had turned into a beautiful chestnut stallion that could immediately sense the tension in the animal on its back. But he was used to the feeling – he'd been trained specifically to deal with it – so he waited with calm resignation for the arrival of its first botched command.

The animal on his back could not sense anything about the animal beneath him, except that it was big, powerful and a long way from the ground.

'There,' said Bert, making a last adjustment to the stirrup leather and the position of Danny's foot. 'Now relax into the saddle and let your heels fall with the weight of your legs. And enjoy. We are only going for a gentle walk.'

Bert mounted Marilyn, as though he did it every day, and with a quick double-clicking sound and the slightest movement of his heels, he walked her slowly forward. With a quick double-clicking sound and a flapping of his legs and elbows, Danny got Charlie to follow and they headed west, skirting the edge of the Canyon wall with the great plain stretching to their left and the cavernous hole a safe distance to their right.

'Hold the reins like I showed you,' said Bert. 'Now move them across your body so that they pull the horse's neck in the direction you want to go.'

Danny steered Charlie further into the plain.

'Loosen your lower back and pelvis; let them move with the natural rhythm of the horse.'

As Charlie looped back to join Marilyn, Danny managed to achieve a degree of relaxation in the middle of his body by tensing everything above and below it.

'The best way is to stop trying,' said Bert. 'And the easiest way to do that is to think about something else, like relative time.'

'Yeah, what's time got to do with anything? You're meant to be explaining how I got up here.'

'That is the problem to understanding your problem. You are stuck in Newton's river of time, carried along with a moving now at the same constant pace, towards the future and away from the past … it is obvious.'

'So it's wrong?'

'There is no constantly flowing river. Like speed through space, speed through time is a relative measurement and just as my now is not your now, so *my second is not your second.*'

'You mean they'll last a different amount of time?'

'If we are in relative motion, the time interval of one second as measured by you will be different to the interval measured by me on an identical clock and we will age at different rates.'

'That can't be right.'

'It sounds crazy, but what are we actually doing when we measure time? If you think about it, we are simply measuring the rate at which events occur in our world based on some form of repetitive cycle. Our first division was the 24-hour cycle we call a day, which we now know is the time it takes the Earth to rotate once on its axis. This was followed by the realisation that the repeating seasons operate over the 365-day cycle we call a year.'

'The time it takes the Earth to orbit the sun.'

'And our notion of months comes from the 12 full moons that appear every year due to its 27½ day orbit around the Earth.'

'What about weeks? And hours?'

'The Egyptians invented the hour by dividing the night into the regular spaced intervals of the twelve main star-groups as they appear above the eastern horizon. It made sense to divide the day into the same number of units, but it need not have been done that way. The week has no astronomical basis and its length is completely arbitrary, ranging from the Ancient Greek's ten days to the Russian introduction of a five-day week in 1929.'

'How did minutes and seconds come about?'

'Again they were human inventions originating with the Babylonians who used a mathematical base of 60 where we now use ten. The point is units of time merely record events that are occurring in regular cycles and the cycles we use are purely a matter of historical accident or choice.'

'The length of a day and a year is fixed by how the Earth moves.'

'Not on Jupiter. It takes her almost 12 of our years to orbit the sun, but her days are less than 10 hours long because her speed of rotation is faster. The Earth's orbital period is actually a little longer than 365 days, which means our clocks are constantly moving out of sync with the seasons. We now leap forward a day every fourth year to make the correction, but when Pope Gregory 13th issued a decree in 1582, and the date suddenly jumped from the 4th of October to the 15th, the peasants rioted in the streets at losing eleven days of their lives and a fortnight's pay.'

Danny laughed, and Charlie pushed his head forward and rumbled his lips, as though joining in.

'It shows the problem people have with the notion of time,' continued Bert. 'We are tied to the Newtonian idea that it is absolute and not affected by anything when it is clear that time is simply the way we measure motion. Galileo introduced the concept of time into mathematics for precisely that reason, since motion is nothing more than the change in position over time. Even Aristotle realised that time was motion, reasoning that if nothing ever moved in the universe we would have no perception of it passing.'

'He's got a point,' said Danny, remembering his vigil at the deserted roadside waiting to be picked up.

'So what is a clock?' Bert smiled as he watched Danny's body sink into the leather of Charlie's saddle. 'Now gently spur him on with your heels.'

A finger-peninsular pointed into the Canyon, and as the edge of the abyss started to turn away and Marilyn started trotting across its base, Danny flapped his legs and elbows and Charlie trotted after her.

'I d-don't th-think I'm do-doing this r-right,' he said, bouncing off the saddle, like a rubber ball attached to a bat by elastic.

'Raise your bottom forward as the horse's back rises and sink down it falls. And try not to pull on the reins all the time … that's it; follow the rhythm of the trot. Now move your right heel back, press it into his flank and click him on.'

131

Charlie's gait was an experimental modern jazz piece for all the rhythm Danny could find in it, but as his leg bounced around, a little further back, Charlie suddenly extended into the rolling dance of a slow canter.

'Yay,' grinned Danny, immediately picking up the rhythm.

'Keep your back straight and those heels down,' said Bert, cantering Marilyn alongside. 'Now you look the part.'

Rocking gently along with the sun on his back, the wind in his face and his trusty steed beneath him, Danny felt the part. He was the Rango Kid, crossing the wild plains that were his home, his eyes narrowed against the mid-morning sun, checking the horizon for Indians.

They cantered across the peninsular, following natural paths between thickets of yellowing brush, until the yawning edge of the Canyon swept back to meet them.

'So come on,' said Bert, slowing the horses back to a walk. 'What is a clock?'

'A clock must measure some form of repeating motion, like the rotating Earth.'

'And for over 400 years we have been finding motions that repeat at faster and faster rates to produce ever more accurate clocks. Pendulum clocks appeared in the seventeenth century after Galileo timed a swinging lamp with his pulse and found that it took the same time to complete every swing. Digital clocks were developed in the 1920's when it was discovered that an electric current passing through a crystal of quartz makes it vibrate very precisely thousands of times every second, allowing us to measure time intervals down to thousandths of a second.'

'I always wondered how they measure one bobsleigh team to be only a thousandth of a second quicker than another.'

'That is nothing. The most accurate clocks today are atomic, measuring time in billionths of a second or *nanoseconds*. Every type of atom has a specific frequency of vibration, and we use caesium atoms vibrating over nine billion times a second to measure time to a ninth of a nanosecond and an accuracy of one second every 350,000 years. But whether it is the Earth's orbit, the mechanical movement of hands around a face or an electronic counter counting the vibrations of an atom, any form of regularly repeating motion is a clock.'

'Okay we can split time up in different ways, but you said my second is not your second when surely it's the same for everyone.'

'You mean two identical clocks will run at the same rate, regardless of how they are moving relative to each other?'

'Yes.'

'Let's take our bouncing ping-pong ball and make that our clock.'

'Can we do that?'

'With you batting it, no. But provided the ball bounces to the same height every time in a regular and repetitive cycle then, yes, it will make a perfectly good clock. Let's imagine we each have a ping-pong clock bouncing, or ticking, once every second. If you are standing at the roadside and at the instant I drive past you in my Beetle at six mph our clocks are synchronised and bounce in time with each other, what would you see after that? We see a body, or event, if it is emitting light or reflecting light back to us, and so the light from one bounce to the next on your clock has a constant distance to travel to reach you and will appear to you to be keeping regular time. When will you time my next bounce to occur on my clock?'

'It'll be eight feet down the road.'

'We do not agree on where the bounce occurred and the fixed speed of light means we are not going to agree on *when* it occurred either. If we count only the bounces after I have passed you, travelling at its fixed speed of one billion feet every second, or *one foot per nanosecond*, the light from the event of the first bounce on my clock is going to take eight nanoseconds to travel that extra eight feet to reach you and will arrive later than the first bounce on your clock. The second bounce on my clock will again be eight feet further down the road and will again be eight nanoseconds later than your second bounce, and so on.'

'The light from your second bounce has to travel sixteen feet so it should be 16 nanoseconds late.'

'By the time you receive my first bounce, eight nanoseconds after your first bounce, my second bounce has already travelled eight feet in that extra time and has only the remaining eight feet left. By the time you receive my second bounce, an extra 16 nanoseconds more time has passed on your clock and the light from my third bounce has already completed sixteen of the extra 24 feet it must travel to reach you. The result is that my clock appears to you to be running slowly by a constant eight nanoseconds every second.'

'What about your watch?'

'It does not matter what type of clock we use or how time intervals are divided, you will see my seconds end eight nanoseconds after each of yours.'

'So your seconds are lasting longer than mine.'

'Which means more time is passing for you.'

'What about from your point of view? You think you're standing still in the Beetle and I'm moving past you at six mph in the opposite direction, so you should see my bounces arrive eight nanoseconds late compared to your clock.'

'From my point of view that is exactly what I see and your clock is running slow. But if I instantaneously start reversing back towards you, what will you observe now that I am approaching at six mph?'

'Each of the bounces on your clock will be eight feet closer than the last. Will they be eight nanoseconds shorter than mine?'

'That's right,' said Bert, reining Marilyn to the left to ease Charlie further from the Canyon edge. 'The light from each successive bounce travels towards you over a distance that is constantly reducing by eight feet every second. My bounces arrive sooner, my clock appears to have speeded up by eight nanoseconds every second compared to yours, and by the time we are back together our clocks are back in sync and read the same time.'

'That means they were ticking at the same rate all the time.'

'Yes, it is called the *Doppler effect*. Remember yesterday when that lorry went past blaring its horn? The sound or pitch of the horn became higher and higher as it approached then lower as it passed and receded away.'

'What's that got to do with the bounces changing on our ping-pong clocks?'

'The two effects are exactly the same and are caused by the wave nature of both light and sound when the source of the wave and the observer are in relative motion. If I am running towards you clicking my fingers once every second, more and more sound waves are arriving into a constantly shortening distance. As we saw with our rope waves, they become squashed together, their wavelengths decrease, more shorter-wavelength waves reach you every second –'

'And their frequency goes up.'

'Exactly. And the pitch of a note of sound is governed by its frequency. Middle C on a piano is a sound wave vibrating at a frequency of 256 Hertz or wave/vibrations per second. Bottom C is 128 Hertz and top C 512.'

'So as the frequency increases the pitch becomes higher.'

'Once I run past you, each click starts moving towards you from further and further away and the sound waves become stretched over a greater distance.'

'Their wavelengths become longer, the frequency goes down and the pitch becomes lower.'

'This change-in-frequency effect applies to all waves, including light. The light waves still move at their fixed speed of 186,000 mps but the number of waves travelling each second can vary depending on the relative motion between the source and the observer. But frequency is really a statement about time.'

'It's how often something's happening.'

'So as the frequency of the light from my Beetle decreases, you see fewer events happen because you are receiving fewer waves in every one of your seconds and everything inside appears to move slowly. When I am moving towards you and their frequency increases, you receive more waves every second and everything appears to run fast.'

'And you speak in a higher pitch, like on the way back from Jupiter.'

'The effect was first explained in 1842 by an Austrian called Christian Doppler, and we have been using it in our police radar guns, which measure the changing frequency of a radio wave as it bounces off other cars to work out their relative speeds, regardless of whether they are approaching or moving away.'

'If the Doppler effect doesn't really change the passage of time then it won't make us age by different amounts.'

'No, but our actual motion through space will.' Bert pulled Marilyn to a stop and Charlie stopped beside her. 'How are you moving?'

'I can't say exactly, but I can say I'm standing still even though from every other point of view I'm moving at a different velocity.'

'Which point of view is right?'

'They're all right.'

'So for every body there are two frames of reference; the stationary frame of their own point of view and the moving frame of an outside point of view. And because the reality of one observer's measurements is as valid as the reality of any other's, both frames are happening *at the same time*. We have two completely different takes on the same set of events that have to fit together and make sense in both the stationary and moving frames.'

Bert spurred Marilyn back into a walk, and not waiting for Danny's flapping legs and elbows routine, Charlie moved off to follow.

'Newton made sense of it with the addition and subtraction of velocities based on his three laws of motion, which enables us to work out the relative speed from any point of view. But that does not work for light. Experiments have shown that light behaves like a normal wave except for the crucial fact that we do not measure its speed to change regardless of the relative motion of any and every observer. That has profound consequences for our understanding of time because it means there cannot be one set interval of time between any two events.'

'What's any of this got to do with me ending-up here, at the top of the Canyon?'

'Because any two events that occur at different times, like you climbing in and out of the Beetle, do not occur at the same place or position in space, even if it seems like the same place to you. When you were bouncing the ping-pong ball in my Beetle, relative to us in the stationary frame two vertical bounces were separated by an interval in time happening at *the same position in space*. But to an observer at the side of the road watching us pass at six mph —'

'It bounced down the road in eight-foot jumps.'

'From *their* stationary frame there was an *extra interval in space* of eight feet between the bounces in our moving frame. To an observer on the North Pole we were moving at 834 mph with 1500 feet between bounces. And the Earth moves in its orbit at an average 30 mps so to an observer on the sun there was an extra 30 miles.'

'The distance the ball travels is different for different observers, I get that.'

'In the same way, two events which for one observer are separated by a distance or interval in space happening at the same time, for another observer will involve an *extra interval in time.*'

'Now you've lost me completely,' said Danny, shifting his bottom along Charlie's saddle in search of a softer spot.

'If you are standing by the roadside with an atomic clock and two lightning bolts strike the ground 20 feet either side of you, your clock will measure the bolts to strike simultaneously after 20 nanoseconds. But if I am driving past in my Beetle and at the moment they strike I am right next to you, what would I see? By the time the light from the bolts reaches the midpoint where you are standing, I have moved on.'

'You'll see the one you're heading towards strike first.'

'If we say I have moved on by one foot, my clock will show an arrival time of 19 nanoseconds for the flash I am approaching and 21 nanoseconds for the flash from which I am receding, separating the events by two nanoseconds. What if, at the same time as the bolts strike, another driver is right next to us going past in the same direction at twice my speed?'

'He'll travel twice as far as you did and see the flash he's approaching after 18 nanoseconds and the other after 22.'

'For him the flashes were four nanoseconds apart. What if there was a fourth observer in line with you at the moment the bolts strike but travelling in his car in the opposite direction?'

'The order of the flashes will be the other way round.'

'And which point of view is right?'

Danny opened his mouth then closed it again. 'We could say me and the Earth are moving and you're standing still in the Beetle.'

'The lightning flashes will be moving with you, so they will still be moving relative to me and I will still see the flash heading towards me strike first. The same can be said for any other observer, so when did the flashes actually occur? What, if any, was the time interval between them?'

'We can't say.'

'Exactly!' said Bert, bouncing off his saddle like Danny in a trot. 'We measure the passage of time by dividing it into intervals using the motion of regular events, but *there is no such thing as a regular event.* Whether it is an up-and-down bounce of your ping-pong clock, two flashes of lightning, an orbit of Io, or you getting in and out of my Beetle, the time-interval between any two events is going to be different for observers in relative motion and we are left with the startling conclusion …'

'Time is relative,' said Danny, staring into the space of the Canyon.

'Our measurement of the passage of time is relative to a body's motion through space.'

13

The horses picked their way along the top of the Canyon, but to Danny the vast hole in the earth was looking less of a monument to the passage of time and more like its grave. Bert's explanation of relative time was beginning to just about nearly make some kind of sense … then again it didn't. 'Isn't the different times we measure for the lightning bolts just the Doppler effect because the light has to travel different distances?'

'No, this applies to the light *inside* the moving frame. Let's replace our ping-pong clocks with light clocks. A light clock consists of a beam of light trapped in a vacuum between two tiny mirrors aligned one foot apart, one directly above the other.'

'Do they have clocks like that?'

'Like ping-pong clocks they are a little impractical, but they would make highly accurate clocks because the beam will keep bouncing the one-foot distance between the mirrors a billion times a second, or once every nanosecond, in a regular cycle. According to Newton, once our light clocks are synchronised they should bounce or tick at the same rate, regardless of out relative motion. But what will you see happen when I drive past you, travelling to your right?'

'It'll take longer for the light showing the picture of your next bounce to reach me.'

'We are not talking about your view of the light waves coming from my Beetle and the Doppler effect because that is not a real change to the passage of time. What must be happening to my actual light beam in my light clock in my moving frame?'

'What do you mean?'

'In your stationary frame at the roadside, the light beam in your clock bounces one foot straight up from the bottom to the top mirror in one nanosecond, and one foot straight down from the top to the bottom mirror in the next nanosecond and so on. But in my moving frame, in the time it takes my beam in my clock to travel from the bottom mirror to the top one, you see both my mirrors travel a little to the right.'

'If your top mirror moves right, your beam won't hit it.'

'It has to. We can argue about where and when an event happens but not the fact that it did happen. From my point of view in the Beetle I am in the stationary frame and my light beam is bouncing straight up and down and definitely hits both mirrors, even though from your point of view I am in a moving frame and those mirrors keep moving to your right.'

'That means your beam isn't bouncing straight up and down.'

'That's right. In order to hit each mirror in my moving frame, my beam travels at an angle and follows a diagonal path. And your point of view is as valid as mine because we are merely observing the same event from each of our perspectives or frames of reference.'

'How can that happen?'

'In the same way as your bouncing ping-pong ball or when you jump up in the aisle of a bus, my light beam is sharing the Beetle's motion and its momentum keeps it moving forward at that same velocity as it bounces between the mirrors.'

'Light has momentum?'

'The point is, my beam's diagonal path in my moving frame must be longer than the straight up and down path followed by your beam in your stationary frame. And because the speed of light is constant for both of us, my beam will take more time to travel that zigzag path.'

'And your clock will bounce more slowly than mine.'

'Exactly! Motion through space causes motion through time to slow down because intervals of time expand in the moving frame to allow for the extra distance travelled. The faster I am moving relative to you, the further my mirrors move between bounces, the longer the diagonal path of my beam, and the greater the difference in time recorded by our clocks. And this is not the Doppler effect; this is the actual behaviour of my clock inside my moving frame.'

'I still see what's happening because of the light waves travelling to me.'

'That's right. If I am moving away from you, the slowing-down-of-time Doppler effect due to the decreased frequency of my light waves simply adds to the real slowing down of time, or *time dilation* taking place in my moving frame.'

'So your clock runs even slower.'

'If I am moving towards you, the speeding-up-of-time Doppler effect due to the increased frequency of those waves will subtract from the real time dilation that is still taking place in my moving frame for that relative speed.'

'Isn't this time dilation just because it's a light clock and the speed of light is fixed?'

'For any good clock the cycle of motion is fixed, and you cannot get more fixed than the speed of light, but any two identical clocks in relative motion do not keep perfect time because there is no perfect time. There is no, one, set interval of time between the same two events on which we can all agree.'

'How do we know it's time itself and not just what we're measuring?'

'We can only go on what we can measure; that is our reality because that is what is happening from our frame of reference. And the reality is, we always measure the speed of light to be the same in both the moving and stationary frames, and that can only happen if the time interval between events expands in the moving frame to give everything *extra time* to cover the extra space through which it is moving in that frame. The result is that less actual time passes in the moving frame compared to the stationary frame, and we have to accept that time is not absolute and its measurement is relative to our motion.'

Danny chewed his bottom lip and stared at the ground swaying between Charlie's ears.

'For centuries this property of time has gone completely unnoticed because the time dilation is tiny at the relative speeds we experience here on Earth. But when you are moving at an appreciable percentage of the speed of light itself ...' Bert left his sentence hanging and pulled Marilyn to a stop.

Charlie stopped beside her, and something in Bert's watery blue eyes made Danny start fidgeting in his saddle. 'What?'

'You see, Danny,' said Bert, pointing into the Canyon, 'the time interval between the events of you climbing into my Beetle down there and getting out up here was not the same for you as it was for me, or anyone else on planet Earth.'

'What are you on about?'

'What time is it?'

Danny fished behind his chaps and pulled his mobile from his jeans. 'Eleven-thirty,' he said, ignoring the persistent lack of signal.

'The sun is too high for 11:30. It is nearly one o'clock.'

Danny looked up. 'It must be running slow ... or something."

'What day is it?'

'You know what day it is.'

'Yes, but I am afraid you do not. It is Friday.'

Danny laughed then tried to look solemn. 'No, Bert. I think *you* will find it's Tuesday.'

'No, Danny.' Bert volleyed it back, but there was no smile at the corners of his mouth and no amused twinkling in his eyes. 'It is 12:47 on Friday July 29th. Your trip to Jupiter took only 40 minutes of Beetle time, and 40 minutes on your mobile in your pocket, but relative to the Earth you have been gone for over three days.'

Charlie pricked-up his ears as a snort sounded and he felt a vibration on his back.

'You're going to have to do better than that.'

Bert dropped his head and spurred Marilyn into a walk, and as Charlie started to follow, a twang shuddered through Danny, like a duff chord through a guitar. 'Well? ... Go on then.'

Bert attempted to pull his cheeks into a smile but settled for pulling up his shoulders. 'You will not believe me, no matter what I say.'

'If you mean that three days have passed and it's now Friday the 29th and not Tuesday the 26th,' said Danny, 'then you're right, I won't believe you.' But it didn't come out as light-heartedly as he intended.

'I am going to have to show you.'

'Then show me.'

Bert nodded slowly, then suddenly crying 'YAH!' he reared Marilyn's front legs off the ground, as her hind legs launched her forward, and started galloping back across the plain.

Deciding that it would be a good idea to follow, Charlie shot off after them. And after throwing his legs into the air and grabbing frantically for the pommel, Danny decided it was best if he went with him.

*

They galloped most of the way back to the Beetle, and Danny had no idea a horse could run so fast. It was like being in a rally car, but there was no racing harness, roll cage or bodywork to protect him, and no familiar controls to a predictable engine. His driving force was a tonne of stretching sinew and contracting muscle, pounding across the uneven ground with only gravity and momentum keeping him on its back.

They walked the last cooling mile in silence, but as soon as they'd dismounted – and Danny managed to get his legs back together – Bert handed him a folded newspaper.

Danny unfolded it into the *Grand Canyon News* and a grainy picture under the news update section immediately caught his eye – an unmistakable VW Beetle suspended beneath tiny parachutes against a vast backdrop of Canyon and sky. 'Hey, we made the paper! It must have been Joe and Billy.'

UFO OVER CANYON

There has been no further sighting or confirmation of the report earlier in the week that a tragic genius NASA scientist and his assistant are engaged in top-secret experiments at the Grand Canyon.

Danny dropped the paper to his side and looked at Bert. 'We-ll, I couldn't help it,' he said, failing to hold back a smile. 'It must've been the adrenalin. And when they started asking why I was driving a car off the top of the Canyon, I didn't know what to say and it just –'

'No, Danny. Look again.' Bert continued to remove the tack from the horses and waited for it to come.

'Eh?' said Danny, spotting the date in the top left corner. He turned the page as though it would correct itself, but it didn't. Halfway down, a headline announced 'Communications Still Down' and underneath, the article began:

> Further disruption to satellite communications is expected this weekend as solar flare activity remains high. But the National Oceanic and Atmospheric Administration report that the effects will be less severe than were experienced in the middle of the week –

Danny stopped reading. 'They've made a mistake. It's Tuesday, not Friday.'

He shuffled through the next few pages, stopping only to glance at the top corner of each – except for noting that, on Thursday, a Martha Platt from Cedar Ridge, aged 62, had been shocked to discover a two-headed piglet in the new-born litter of her pet pig, Bessie, aged four. But the photograph looked faked to him.

Bert gave Marilyn and Charlie a smack on their rears, and as they pranced away to graze on the open plain, he threw another newspaper to Danny and started to set up camp.

It was the *New York Times*, a very different newspaper but with the same date, Friday July 29th, jumping out from the top of every page and references to events throughout a week that, as far as Danny was concerned, had only just started. 'This can't be right.'

'Here,' said Bert, opening one of the chairs and passing Danny the briefcase with the satellite dish spinning above its lid. 'You should be able to get on line; there have been less disruptions today.'

Danny sat onto the chair, until every site confirmed the date as the 29th and he jumped back up.

'Here,' said Bert, from the open passenger door of the Beetle, waving Danny to join him and leaving as soon as he did.

Danny lowered himself into his seat as the last bars of a country ballad sounded from the speakers followed by a chorus of voices singing the rising chime, *'K-N-A-G on Nine-ty point Three'*, and one chirpy one announcing, '*I'm Bob Singer and this is Grand Canyon Radio coming to you from beautiful downtown Grand Canyon Village. Stay tuned and after the news we'll be talking to Martha Platt and finding out about her strange discovery.*'

'*I opened the door to the barn and there they were, staring back at me,*' said Martha. Then Bob leapt back. '*It's 1.30 on Friday, July the 29th and here's Judy with the lunchtime news and weather.*'

'Press key number two,' called out Bert, from a pile of folded furniture.

Stilted English replaced Singer's laid-back American, but when the BBC World Service confirmed the date on their mid-evening news, Danny's frown turned into a smile and he climbed back out of the Beetle. 'This is another set up, isn't it?' he said, poking the old man in the ribs and watching for a tell-tale sign.

'I wish it were; it is my fault that this has happened to you.' Bert lifted the tabletop onto its edge and opened its legs either side of his own. 'I know it is hard to accept, but just as you cannot travel through time without travelling through space,' he intoned, like a preacher behind a pulpit, 'so you cannot travel through space without travelling through time.'

'Don't start all that physics mumbo-jumbo.' Danny span round in a circle as though looking for hidden cameras. 'There's no way that can happen and you know it. I was only in your Beetle for an hour.'

'I know only too well that it can.'

'How do you know?'

'Because that's what happened to me!'

It was the first time Bert had raised his voice in anything other than excitement or joy, and he turned away and dropped his body into the chair and his head into his hands.

'I am a physicist. Or at least I was,' he said, lacing his fingers through his thin white hair and slowly massaging his scalp. 'I had retired and with time on my hands I built the Beetle. It only started as an exercise to help a friend understand my theory of relativity. I did not know anything like this would happen. On the test run I spent over three hours playing out different scenarios, moving relative to different bodies like stars and neutrinos. By the time I was done over 50 years had passed on Earth.'

Bert lifted his head and twisted his mouth into a smile, but his eyes didn't join in. 'Like you, at first I could not accept it, even with the evidence of decades of scientific advancement all around me. I climbed into my Beetle at 2:15 in the afternoon on April the 12th 1955 in my driveway in Princeton, New Jersey. I climbed out at 6:23 in the morning on November the 20th 2011 here at the Grand Canyon. My friends, my house, my entire life was gone and I was trapped here.'

His head fell again then rose to flash a brief but real smile. 'It is quite something to read your own obituary. They could not say I had gone missing because they suspected I had defected to the Russians, and so would everyone else. They, by the way, are the FBI. I had been outspoken against the US government for some time, and I knew Hoover was trying to find something to use against me. Like many intellectuals in the 1950's I was denounced by Senator McCarthy and his committee of communist witch-hunters, so when I disappeared it was the natural assumption. The Russians denied it of course, but then they figured they would. And though I had more or less retired from public life, it was only a matter of time before the truth came out, so to avoid a national embarrassment they killed me off. Apparently I died peacefully in my sleep.'

Bert looked earnestly up at Danny. 'But how could it possibly have happened? I knew that the time dilation could occur under my theory, but as for the mechanics –' he snorted and shook his head. 'Anyway, it became my mission to find out. It is the one thing that has kept me going ... without it I would have given up on everything.'

Danny had no idea what to say to most of Bert's explanation, but he'd finally picked up on something he did. 'What are you talking about, *your* theory?'

'Did I not mention that?' said Bert, with an innocent air but a guilty look. He paused to swallow and change it to an acutely embarrassed one. 'You see, the thing is ... well ... I am Albert Einstein.'

Danny started laughing. He wagged his finger at Bert and tilted his head. 'You had me going there for a moment.'

'Is it that unbelievable?'

'It's almost as unbelievable as that question.'

'How did you get up here, Danny?'

Danny stopped laughing. He stared at Bert until he was looking straight through him then leapt straight at him, his hand extending to grab the Motorola out of the briefcase. 'I got to make a call,' he said, extending the aerial and tapping at the keypad as he passed the Beetle.

'I know you cannot believe me,' Bert called after him. 'But I am close to understanding how it could have happened to us both.'

Danny scrambled up the loose limestone wall at the back of the garage and dropped to his bottom at the top. Pulling his knees to his chest and the brow of his Stetson down over his eyes, he pressed the phone to the side of his head.

'Mum!'

'Danny! I was wondering when you were going to call. Too busy having fun?'

'Is it Friday?'

'Sorry darling?'

'I said, is it Friday?'

A sigh sounded through the earpiece.

'What are you going to come up with this time? Danny, I don't care what it is, you've only got –'

'Is it Tuesday then?'

'I wasn't going to tell you until you got back, but I've been thinking and a one-week camp is not enough to –'

'Mum!' Danny shouted over her. 'Is today Tuesday the 26th?'

'No, today is not Tuesday the 26th.'

'It's not past midnight there, is it?' But he knew it couldn't be.

'It's the 29th and you only have one full day left so there's no point griping now. Look we'll talk when you get here. How's it going at the ranch? ... Danny?'

'Hm? Oh. Yeah.'

His mum laughed. 'Yeah what? We're having a wonderful time here with Great Uncle Gil but we are missing you. Alice has drawn a beautiful picture of you riding a horse, haven't you darling. Hang on ... She wants to tell you herself.'

14

Fluffy clouds hovered on the flat horizon of the Canyon's north rim, like grazing sheep, and as a straggler drifted across the middle towards them, its shadow undulating over the mountains below the brim of his Stetson, Danny picked up the Motorola and his empty plate and climbed down from his limestone perch.

Bert had set up camp with the fire down on the sandy plain, and the table and cool-box on the rock platform in front of the Beetle. Anchored by the roof rack and two telescopic poles, the parachute canopy billowed above them, shading the old man as he hunched over a fat, metal cylinder and turned a screwdriver in the slot in its side.

'Thanks for the bacon and eggs … Bert.' Danny had discovered them steaming on the garage wall behind him, but Bert was already back by the fire so he'd eaten and said nothing.

Bert looked up and smiled. 'I thought you might be hungry.'

'Yes. I was.'

They held each other's eyes then both looked away.

'I ran the MFDP on my briefcase computer,' said Bert, tightening a screw near the top of the slot. 'The Beetle's electrical fault has returned and we will not be able to get going until I fix and fit this back.'

'What is that?'

'It is the Beetle's dynamo or electric generator. Inside the main body spinning magnets create an electric current in a coil of wire and the electricity is transferred to the battery along –'

'Magnets create electricity?' said Danny, leaning over the table and peering into the opening in the generator.

'Yes. Surprising isn't it? Both the Ancient Greeks and Chinese knew about magnetism and electricity, but it was the nineteenth century before experiments showed that an electric current flowing through a wire causes a compass needle to deflect like a magnet, and two parallel wires will attract or repel each other depending on which way an electric current flows through them. The two forces were clearly connected in some way, and a self-taught English scientist called Michael Faraday showed how. He visualised the space around a magnet being filled with invisible lines of force. He then proved it with nothing more than a pinch of iron filings and a piece of wax paper.'

As Bert slipped comfortably back into his physics, like an old pair of his shorts, Danny lowered his bottom gingerly to his chair, happy to settle there with him. 'We did that at school. You sprinkle the filings around a bar magnet and those lines appear.'

Bert placed his fingers deep into the slot in the generator, looking off to the side as he rummaged around inside, like a vet with his hand up a cow. 'The iron filings trace out the magnetic lines of force as looping butterfly wings on each side of the body of the magnet, forming an invisible field of influence where the force can be felt.'

'A force field.'

'That got Faraday thinking and in 1831 he passed an electric current through a copper wire suspended above a magnet. This made the wire behave like a second magnet, and as it was alternately attracted and repelled by the first, it moved continuously round and round to produce the world's first electric motor. Reversing the situation, when he moved a magnet around a copper wire he induced an electric current to flow in the wire to produce the world's first electricity generator.'

Bert repositioned his fingers in the generator. 'The electricity was only produced when the magnet moved, and like the kinetic and gravitational energy of a falling hammer, as the amount of electricity went up, the available magnetism went down by a proportional amount. In other words, magnetism and electricity are not separate forces; they are different aspects of a single *electromagnetic force*. According to Faraday, this combined electromagnetic field could support wavelike disturbances on its own and light was a vibration or wave within the field. There was no need for the Ether and this was 50 years before the Michelson/Morley experiment proved it.'

Having had no idea where it was heading when they started their conversation, Danny was surprised to discover it led back to light and the mystery of its fixed speed.

'But if they were to be accepted, Faraday's theories needed a mathematical framework to support them. That brings us back to James Clerk Maxwell and the four electromagnetic field equations he discovered in 1864, which explain every magnetic and electrical effect and the true nature of light.'

'Light's an electromagnetic wave moving at 186,000 mps.'

'That's right. Maxwell's equations show that a varying electric and magnetic field can oscillate through space as an electromagnetic disturbance or wave. The speed of these electromagnetic waves in a vacuum is a constant, which Maxwell labelled with a small c from the Latin *celeris* meaning swift. This constant c is derived mathematically from two other fundamental constants in Nature: the strength of the electric force between charged particles and the strength of the magnetic force between magnets. The measurement of these values between any two particles and any two magnets never changes, which basically means when you put the numbers into his equations, like magic, up pops the speed of light as the value of c.'

'So Maxwell wasn't looking for an explanation of light?'

'No. He was trying to find a mathematical explanation for Faraday's electromagnetic effects. It was a surprise to everybody when the speed of light turned up.' Bert removed his fingers and peered into the generator. 'The point is,' he said, shoving them back in, 'Maxwell's constant c for the speed of light is not just a random number chosen to fit the facts; it is a logical and direct mathematical consequence of the nature of our universe.'

'Okay the maths says the speed of all light is always the same, but how –'

'The speed of light is always 186,000 mps in a vacuum, but the equations show that it is slightly reduced when moving through denser materials such as air, glass and water. They also show that, just as there can be any frequency sound wave even though we cannot hear them all, depending on its energy there can be any frequency electromagnetic light wave even though we cannot see them all. The different colours of the rainbow we see with our eyes, which combine to form white light, consist of electromagnetic waves vibrating at different frequencies around 500 trillion waves or cycles per second, with red at the lower end of the spectrum and violet at the top.'

'You mean each colour is a different frequency light wave in the same way the notes on a piano are different frequency sound waves?'

'Yes, but the spectrum of light that we can see with our eyes is only a tiny bandwidth lying roughly in the middle of the entire electromagnetic spectrum of all possible frequency light waves. Infrared, Micro and Radio waves lie below the red end of the visible spectrum, with wavelengths ranging from thousandths of an inch to miles. Ultraviolet, X and Gamma rays lie above the violet end, with wavelengths becoming smaller than an atom and frequencies as high as a million trillion cycles per second. According to Maxwell's equations, they are all electromagnetic waves and they all move at the constant speed of 186,000 mps through the vacuum of space, relative to everything.'

'Yeah, but *how* do they do it?'

'Nobody knew. The science community was trying to reconcile both Maxwell and the Michelson/Morley experiments with Newton's laws, but this approach meant that the Ether had to exist as the medium through which light travels at its constant speed. Even Maxwell was not sure how light moved under his equations and clung to the idea of the Ether. Yet the real explanation lay there all along, waiting to be found, but no one could see it ... except me.'

They looked at each other from under hooded brows, like boxers at the sound of the bell.

'Albert Einstein?' said Danny.

Bert nodded and pulled his fingers out of the generator. 'It is no good. Do you see that metal wire down there?' he said, pointing into the hole with his screwdriver. 'I have to clip the end over that little copper contact and secure it with this screw, but my fingers are not as adept as they used to be.'

'I can do it.' Danny lowered his fingers into the slot and looked up, watching the sunlight ripple through waves of parachute silk, as though reflecting off water. 'So how old are you?'

'One hundred and thirty-seven.'

Danny laughed.

'I have only lived through 81 of them but I was born in Germany in 1879.'

'It's clipped on,' said Danny, heading back in with the little screw.

'Ah, to be your age again,' said Bert, pulling two bottles from the cool-box and relaxing back in his chair. 'I was sixteen, only a little older than you, when I first started thinking about the nature of light. I wondered what would happen if you chased after a light beam. And if you could catch up and ride along with it, like a surfer on a water wave, what would you see?' He found his pipe in the top pocket of his black waistcoat and popped it into his mouth. 'According to Newton's second law, as long as we keep applying sufficient force from our rocket engines, we can accelerate Spaceship Beetle to match the wave's velocity, which means relative to us –'

'It'd be standing still.'

'There would be no relative motion and nothing would be waving. Now suppose the beam of light we are riding came from your Town Hall clock as it struck midday. What time will you see when you look back?'

'Won't it depend on how far away it is?'

'No. You are surfing on a light beam at the speed of light, so no matter how far you travel you cannot catch the beams that left before yours, and the beams showing the picture of the clock at all later times cannot catch up with you.'

'Then the clock will always show noon,'

'What would you see if you were moving at the speed of light in Spaceship Beetle, with the window screens raised and the roof light on, and you looked at yourself in the rear view mirror?'

'I'd see my face.'

'We know that now because the speed of light is constant for all observers regardless of their motion. But back in 1895 the generally held belief needed to explain Maxwell's constant was that light's fixed speed must be relative to the stationary Ether. Spaceship Beetle is already travelling at that speed through the Ether, so the light from the bulb, which is situated behind your head, can never catch up with the mirror and bounce back and you would not be able to see your face.'

'That doesn't sound right.'

'That's what I thought. If that happened, we would have to abandon the principle of relativity because you would be able to tell you are moving without lowering the window screens and referring to some relative movement outside.'

'When my face disappeared I'd know I was travelling at the speed of light.'

'Not only could you tell you were moving, you would know what speed you were doing. To me that was wrong. Even though back then we did not know for sure, I believed that everything is moving and we can only measure motion by comparing one body's position to another. That is the reality of our world, not imagining everything to be gliding through some ethereal substance no one can measure to exist.'

'But light doesn't behave in a relative way. Its speed is the same for everybody.'

'I spent ten years thinking about that. Then in 1905 I realised what is really happening in a beam of light and why we can only measure it to be travelling at Maxwell's constant c.' Bert stopped to beam out an especially big grin, yet somehow hang onto his pipe.

'Go on then,' said Danny, dropping the screwdriver onto the table and grabbing his bottle.

'The principle of relativity says that you can only measure your motion relative to another body. So if you are moving at a constant velocity in Spaceship Beetle with the screens raised ...'

'I won't be able to work out I'm moving without lowering them.' What does that mean?'

'It means everything behaves as though I'm standing still.'

'Which means we could have a game of table tennis or repeat the Michelson/Morley experiment in Spaceship Beetle and the physics would be the same as on the Earth or anywhere else at any other constant velocity. That is the full meaning of the principle of relativity: *the laws of physics are not affected by constant-velocity motion and do not favour any specific point of view or frame of reference.*'

'Light's point of view is special.'

'Maxwell's constant seemed to go against this principle so one or other of them had to be wrong, but which? I felt the principle of

153

relativity was far too sensible to abandon, and Maxwell's equations combine and explain all electric and magnetic effects with such simple beauty, they had to be true and I did not want to abandon them either. Then the revelation hit me; I did not need to choose. Both ideas are right and with no need for a medium like the Ether.'

'How can they both be right?'

'Because light *does* follow the principle of relativity that the laws of physics must be the same for all observers. Remember that Maxwell's constant c is a mathematical consequence of two other constants in Nature, the strength of the electric force between charged particles and the strength of the magnetic force between magnets.'

'Their values are always the same.'

'They are physical laws which, according to the principle of relativity, all observers will measure to be the same no matter what their velocity. So if we were to measure different values for the speed of light, the values of the other two constants would have changed, the results of set experiments would be different and we would know we are moving without reference to an outside body.'

'So everyone must come up with c for the speed of light.'

'The speed of light has to be fixed, which means Maxwell's constant is a new *universal law of physics* and must be due to the nature of light itself. Light waves are not pulses of energy moving through anything, they are the result of the dual faces of electromagnetism.'

'I thought all waves have to move through something.'

'Electromagnetic waves are not like other waves. When you turn on a light bulb, you are basically jiggling an electric charge up and down, like I jiggled the rope. This changing or *moving* electric field creates a changing or moving magnetic field, which in turn creates another moving electric field, which creates another moving magnetic field, on and on in a continuous process which spreads out in all directions as an electromagnetic disturbance or light wave.'

'You mean the light wave powers itself along?'

'And such a self-sustaining wave has no need for this immeasurable medium called the Ether. Ha, ha! That's it! We can catch and ride a water or sound wave by travelling through the medium at the same velocity as the wave, but we can never catch and ride a light wave because a light wave cannot stand still. A magnetic field has to be moving to generate an electric field and an electric field has to be moving to generate a magnetic field. If nothing is moving relative to us, there can be no light wave at all.'

Danny started nodding then stopped. 'I still don't get why the speed of light's the same for everyone no matter how they're moving.'

'Electricity and magnetism are the two faces of the same electromagnetic force and all motion is relative. One observer's stationary electric field is another observer's moving magnetic field, and vice versa, which means for every observer a new magnetic or electric field is always being generated and is moving away *from that point* at the speed of light.'

'Oh-h. A new light wave begins wherever and whenever you measure it.'

'Exactly. An electromagnetic wave is not simply a wave; it is a physical process that continuously replicates itself at c into a wave. Every jump of a magnetic field between positive and negative poles creates a moving electric wave at that position. And every jump between a positive and negative electric charge creates a moving magnetic wave at that position. These two wave fronts combine to form a single electromagnetic wave of light that always travels at 186,000 mps in a vacuum.'

Bert dropped his palms flat to the table. 'That is the real explanation that had lain hidden in Maxwell's equations for nearly 40 years. Every body must measure their own and each other's light to be moving at c in order for the laws of physics to remain the same for all observers and for there to be no special point of view or frame of reference. It is just that we must now include and accept Maxwell's constant as a new universal law.'

'What about the addition and subtraction of velocities?'

'Newton's formulation must be wrong because it conflicts with our new law of Nature and does not work for light. Those equations are based on Newton's own laws of motion so they cannot be right, as is his universal law of gravitation, so it must be wrong as well. Though I could not see its full extent, sitting at my desk in the Swiss Patent Office in 1905, I realised that Maxwell's constant suggested a link between space and time beyond anything we had imagined. That magical speed of 186,000 mps was going to lead us to an understanding of the nature and origin of the universe itself!'

As Bert sat happily wrapped in his memories, Danny tried to keep from remembering his. Since the conversation with his mum, he'd pushed the evidence to the back of his mind and he was fighting a running battle to keep it there. He had no idea that it was about to make a surprise attack.

'It appears we have company,' said Bert, rising from his chair and shielding his eyes as he squinted across the plain.

Together they watched a small car, weaving and bouncing towards them in a cloud of dust, turn into a large 4 x 4 with a rack of blue lights on top of its cab and *Grand Canyon Sheriff's Department* emblazoned in gold down its sides.

15

With its front wheels on opposite lock, the 4 x 4 ploughed to a sideways stop 20 yards short of the camp. The dust clouds settled but the officer did not climb out, and because his image through the glass was mostly obscured by outside images reflecting back off it, it was difficult to see why – although Danny thought he spotted a pair of binoculars and a radio taking turns to cover his face.

'It's Sheriff Winton,' hissed Bert, two posses of fingers searching his body for his pipe.

The Sheriff scaled down the outside of his wagon, adjusted his mirrored sunglasses and the tilt of his cowboy hat, and swaggered over, as though between sets at the gym. His uniform was tan-coloured and military in style, decorated with pips and insignia, like medals on a hero, with a utility belt holding up sharply pressed trousers and loaded as though going to war. He was in his mid forties with a large jaw jutting out from a fat head, but like a folded pipe cleaner, his body seemed to have one, overall thinness, and as Bert stuttered forward to greet him he found he was almost as tall.

'How the hell did you boys get out here?' drawled the Sheriff, hooking his thumbs into his belt as he surveyed the strange-looking Beetle, strange-looking cowboys and strange-looking briefcase with a satellite dish on its lid.

'Good afternoon, Sheriff,' said Bert, almost bowing. 'Beautiful weather we are having. Yes, my colleague and I were just saying the same thing. Ha! It was not easy.'

The Sheriff found this neither amusing nor informative and having done the good-guy part of his routine for long enough, he felt it was time to move on to his favourite part.

'What day is it?' said Danny, surprising the Sheriff into an immediate answer.

'It's Friday. Don't you know what day it is, boy?'

'It can't be Friday,' cried Danny, waving his arms and spinning away from the table. 'It was only Tuesday a couple of hours ago!'

The Sheriff jumped back and his left hand hovered at his hip, until he remembered his gun sat on the other side and his right took over. 'Have you been smoking something, boy?'

'Like, I wish,' muttered Danny, folding his arms and falling back against the Beetle.

'Do not mind him Sheriff,' said Bert. 'He has forgotten his mother's birthday.'

The Sheriff kept looking Danny up and down. 'Are you that NASA scientist fella?' he said, brushing his finger under his nose.

Danny's eyes glared beneath the brow of his Stetson as he watched the Sheriff examine him.

'We heard the news had got out,' said Bert, moving to stand beside the Sheriff – and not entirely straightening his legs on getting there. He lowered his voice and leaned in. 'My apologies, Sheriff. You know how temperamental these genius-types can be. But he is vital to our research here and our national security in these troubled times, so we put up with the odd tantrum.'

The Sheriff nodded slowly, his eyes frisking the charmingly-short old man before snapping back and barking, 'It don't mean you know everything, boy', as though Danny had been in on the conversation. He looked off to nowhere then back again. 'Okay, let's see your driver's permit,' he said, beckoning Danny to bring it over and do it quickly.

Danny's arms uncrossed and he stood off the Beetle and looked to Bert.

'Here are our official permits,' said Bert, pulling a thick wad of papers from the side pocket of his jacket. He passed them over and his hand continued up to the Sheriff's shoulder. 'I hope you appreciate the confidential nature of our mission here, Sheriff, and that we can rely on your discretion. In fact, sir, we may need your help.' He glanced around, as though checking nobody was eavesdropping, and started steering the Sheriff away from Danny. 'This information is only for high ranking officials, like yourself, but we have been seconded from NASA to Military Intelligence and our experiments are classified level nine ...'

Danny watched Bert lead the Sheriff back to his 4 x 4, with the Sheriff looking between the forms and Bert, and looking less and less at the forms. They chatted for a while at the side of the cab, Bert nodding and appearing to say 'yes' a lot as he opened the big door and half-guided and half-shoved the Sheriff up into it. With a final handshake through the open window, the Sheriff gunned the engine and pulled away.

'Thank goodness, we got rid of him,' said Bert, lowering himself back into his chair. 'He will be visiting us at the Kennedy Space Centre next month, but by then I will be long gone.'

'What is this expedition of yours?'

'I am going to do it again.'

'Do what?'

'Jump to Earth's future.' Bert watched Danny's eyes flip though a catalogue of emotions. 'I think you know how I feel. You will not be at peace until you understand.'

'I still can't believe it, let alone understand,' said Danny, dropping onto his chair and wincing as his bottom took the impact.

'You are so close, Danny.'

'I'm flying back to Phoenix tomorrow night!' Danny dropped his head. Twenty-four hours ago he'd have been dancing if he'd known three days of the week were simply going to vanish. Now, like one of Pope Gregory's peasants, it felt like they'd been stolen.

'Have you spoken to your mother?'

'Yeah, but I didn't say anything. Not that she'd have listened; I could tell she was busy doing something else, as usual.'

'Yes, it must be difficult for her; holding down a fulltime job, running the house, doing the shopping, cooking all the meals and looking after little Alice. It must be a great comfort having you there to help. The problem is, we now have only one day to get to the bottom of how this can have happened to us both.'

Danny decided to go with the change of subject. 'Is it to do with the theory of relativity?'

'My theory of relativity comes in two parts: the *special* and the *general* theory. But yes, it has everything to do with them, and you have nearly all the information you need to understand why. Using

159

my principle of relativity that the laws of physics are the same for all freely moving observers, our basic understanding of relative motion remains in tact, without recourse to either Newton's laws or the Ether, provided we accept Maxwell's constant for the speed of light as a new universal law.'

'Are all Newton's theories wrong?'

'Let's look at what must happen when different observer's measure the speed of light. Let's imagine you are deep in empty space and you switch on a light bulb. What will you see from your stationary frame of reference?'

'There's nothing to see if I'm in empty space.'

'I mean the light will travel an equal distance in all directions around you to form a sphere of light growing in radius by 186,000 miles every second with you at its centre. What would I see if I am cruising past in Spaceship Beetle and at the instant you turn on your light, I am right there beside you?'

'You'd see me.'

Bert's jaw paused in mid fall. 'According to Newton's formulation of the addition and subtraction of velocities, I should measure the speed of the wave front in the direction I am moving to be less than c, and the speed of the front moving in the opposite direction to be more than c. For me, the wave fronts from your bulb are travelling at different speeds and they will not form a sphere.'

'But the speed of light is fixed.'

'Exactly. That little c in Maxwell's equations means I have to measure the same speed for your light as you do. That means I have to see and measure a sphere of *your* light to be spreading out around *me* at 186,000 mps with me at its centre as well.'

Danny laughed. 'We can't both be at the centre of the sphere from my bulb.'

'We cannot under Newton, but how do we measure speed?'

'It's the distance travelled divided by the time.'

'So if we travel 100 miles and it takes us four hours then our speed was 25 mph. In the same way, if we travel at a fixed speed of 25 mph for four hours then we will travel 100 miles. In other words, a distance measurement can be found by multiplying the speed by the elapsed time. This makes light the perfect universal tape measure because we can swap any measurement of distance for an equivalent measurement of the time light takes to travel that distance.'

'Oh yeah, because it travels at the same speed for everyone.'

'If it takes a beam of light one second to reach another body, we know it is 186,000 miles or one light-second away. If the beam takes one minute, the body is 11.2 million miles or one light-minute away. And if it takes one year, it is 5880 billion miles or one light-year away. And because space is so vast, rather than dealing with huge numbers, we say the Moon is 1.4 light-seconds from the Earth, the sun an average distance of 8.3 light-minutes, Jupiter 35 light-minutes and Proxima Centauri 4.2 light-years.'

'And a foot is equal to one light-nanosecond.'

'That's right. In fact we can measure time more accurately then we can measure distance and the metre, the international standard unit of measurement, is now defined as the distance light travels in 0.000000000335640952 seconds as measured by a caesium beam atomic clock. But the point is, space and time are interchangeable and using the speed of light we can convert one into the other. Now we can see that in order for the speed of light to remain constant for all observers our measurements of both space and time must be relative. It is simple maths.'

Like standing still, to Danny there was no such thing.

'Come on,' said Bert, rising from his chair. 'It is easier to see from a spaceship.'

*

The Beetle settled under the twinkling light from the LED star field, and though the headrest argued for its removal, Danny settled with his Stetson in his seat.

'Here you are in the stationary frame in Spaceship Beetle,' said Bert, 'all alone out deep in empty space. What if I come shooting past in Spaceship Beetle II and you measure me to be travelling at 90 percent of c, or 161,000 mps, relative to you?'

'Then from your point of view I'm doing 186,000 mps in the other direction, the same speed as the computer showed on the trip to Jupiter. Where's Spaceship Beetle II?'

'The program is corrupted and I have not had the chance to fix it. Now if a third Spaceship Beetle rockets past both of us, what would we each measure her speed to be?'

'It'll be different for each of us because she'll be going in opposite directions relative to each of us.'

'Under Newton, Spaceship Beetle III is travelling different distances through space but in the *same interval of absolute time*. So in our speed equation, if we are each dividing different distance measurements by the same interval of time, we cannot both measure her to be doing the same speed. But when it comes to light, that c in Maxwell's equations demands that our measurements of space and time intervals do produce a fixed speed. The only way that can happen for all observers, even though their distance measurements differ -'

'Oh-h, their time measurements have to be different as well.'

'Exactly. We have swapped Newton's absolute time for an absolute speed for light where observers must measure it to be moving different distances in *different intervals of time*. That is what the maths is saying must happen, and as we have seen, once you remove the blindfold and think about how we measure time, it is clear that the maths is right. There is no universal rate of time because there is no such thing as a simultaneous event and no set interval of time between any two events. Motion through space causes time intervals to expand and motion through time to slow down.'

'I still don't really get that.'

'That is because you do not have the full picture. So, you are sitting in the stationary frame in Spaceship Beetle with the beam in your light clock bouncing straight up once every nanosecond, and then straight down. What will you see happening to my light beam in my light clock in Spaceship Beetle II as I rocket past in a moving frame at 90 percent of c?'

'It's doing an up-and-down zigzag as it chases your mirrors through space.'

'In your stationary frame you have to see your beam in your light clock bounce one foot straight up or down, but in that same interval of one nanosecond on your clock, you have to see *my* zigzagging beam travel one foot in one nanosecond on *my* clock in my moving frame. That can only happen if the passage of time slows down in the moving frame so that the light has extra time to cover the extra distance it is moving through space relative to the stationary frame.'

'If I measure your time to have slowed down, and from your stationary frame you measure mine to have slowed, yet both our times are normal in our stationary frames, is it really happening?'

'What is normal? There is no fixed rate for the passage of time, but because every frame of reference measures light to move one foot in one nanosecond, the passage of time always appears *normal* and every observer can say they are in the stationary frame.'

'That means different observers are going to have to come up with different rates for my time depending on how fast I'm moving and how much extra space my light has to travel from their point of view.'

'That's right. Time is relative to the speed of a body through space,' grinned Bert. 'In fact, we can measure the amount of time dilation using a simple right-angled triangle and an equation discovered by a Greek mathematician called Pythagoras in 500 BC.'

'Ah, Pythagoras.' Danny nodded then stopped. 'What did he do again?'

'Pythagoras discovered that the square of the diagonal side, or hypotenuse, on a right-angled triangle is equal to the sum of the squares of the other two sides. So if the vertical line in a right-angled triangle is three feet then squared it is nine feet. If the horizontal line is four feet then squared it is 16 feet, which means the diagonal hypotenuse is the square root of 25 and is five feet long. In the same way, we can make a right-angled triangle for every bounce or tick of the beam in a moving light clock. The vertical line is the one-foot distance between the mirrors and the horizontal line is the distance travelled by the mirrors in the direction of motion, which will vary depending on the relative speed. By adding together the square of those two figures and taking the square root –'

'We can work out the length of the diagonal path of the beam.'

'And because light travels one-foot per nanosecond for all observers, we can work out how much longer *in time* it will take the beam to travel the extra distance to its next bounce, and how much slower time will pass for a clock moving relative to us at any speed. If you are moving at half the speed of light relative to me, the time dilation equation says for every hour that passes for you, one hour and nine minutes will pass for me. At around 90 percent of the speed of light it will take the beam in your clock twice as long to catch up with each mirror and for every hour that passes for you, two hours will pass for me. As you approach the speed of light itself, each tiny increase in speed results in a huge increase in the time it takes your light-beam to catch the next mirror and tick. At 99.995 percent of the speed of light, for every hour that passes for you over 100 will pass for me; at 99.99995 percent over 1000 hours will pass for me and so on. And if you were able to reach the speed of light itself, like light-surfing away from your Town Hall clock, your beam would never catch up with the mirrors, the next tick would never arrive and –'

'Time would stop completely.'

'Nowadays, using machines called particle accelerators, we create high-speed time dilation whenever we like. The most powerful is the LHC, the Large Hadron Collider at CERN, the Centre for European Nuclear Research in Switzerland.'

'The one that broke down?'

'It was to be expected from the most complex machine ever built. In a 16.8-mile circular tunnel, buried over 200 feet underground, beams of atomic particles are accelerated in opposite directions by powerful magnetic fields to speeds approaching the speed of light. The beams are then aligned to smash together at enormous energies and we study atomic structure in the shower of particles created by the collision. Because every type of atomic particle vibrates at a specific frequency, we can compare particles in the accelerator with identical particles at rest in the laboratory, and every particle's frequency has always been found to decrease by the exact amount predicted by my time dilation equation.'

Danny thought for a moment. 'Okay then, if my light clock really does slow down in my moving frame then it must be running slowly in my stationary frame at the same time, so how come I don't notice it?'

'Because everything in your moving frame is travelling through extra space and it takes extra time for any two consecutive events to occur. The beam in your light clock ticks less often; your heartbeat slows; the synapses in your brain do not fire as quickly; and the very atoms of which you are made vibrate at a lower frequency. You are viewing and processing your slow-moving world with a slow-working brain, everything appears normal and you think you are standing still.

'Is that what happened to us?'

'As I said, to understand what is going on you need the full picture and we have not looked at what the relativity of time means for our understanding of space.'

'What do you mean?'

'Just as the fixed speed of light means my second is not your second, so it means *my inch is not your inch.*'

16

Leaving the back of his Stetson on the headrest and the front tilting down over his eyes, Danny groaned and sank into his seat. 'This is getting too weird. How can your inches be a different length from mine?'

'If our measurement of time is relative,' said Bert, 'then our measurement of space must also be relative.'

'It is relative. Different observers measure a body to move different distances.'

'What does that mean? Newton believed in absolute time and absolute space. For him space acted like an enormous grid against which you could, in theory, give everything an exact position and measure the exact distance between events.'

'Wouldn't that be an Absolute Position?'

'That's what Newton's critics argued. They said space is nothing, not something, so how can you make measurements against it? And if you could, we would be able to gauge the motion of a single body all on its own in the universe, yet it is precisely because we cannot that we can say we are standing still and all motion is relative in the first place.'

'So what are you saying?'

'Motion causes intervals of time to expand in the moving frame, and it causes intervals of space to shrink, or contract, in the direction of that motion. According to the maths that is what must happen because all observers' time measurements have to divide into their distance measurements to produce c as the speed for all light. It sounds crazy but that has more to do with it being so alien to your experience than the fact it is difficult to follow the logic.'

Danny didn't think it was.

'In fact, not long after the Michelson/Morley experiment, an Irishman called George Fitzgerald put forward the idea of *length-*

contraction as a possible explanation for the unexpected result. He suggested that the beam travelling in the direction of the Earth's orbit *was* slowed down by the Ether wind, but the extra journey time was cancelled by the Ether shrinking that head-on path so that it took the same time to travel as the right-angled beam.'

'You can't just say that.'

'It was already known that all objects moving through our atmosphere contract a little due to the force of the air pressing on them, so maybe the Ether was having a similar effect.'

'Was the Ether supposed to have any affect on the right-angled path?'

'Everything contracts in the direction of motion so the right-angled path would be thinner but not shorter.'

'Oh yeah.'

'Then in 1895 a German called Heinrich Lorentz showed that Maxwell's equations predict an increase in the force between the atoms in a moving body. This stronger force squeezes the atoms together and causes the body to contract in the direction of motion, and he came up with an equation that showed the degree of length-contraction for any given speed. As it could not be proved experimentally it was not given much attention, but I realised that this Fitzgerald/Lorentz length-contraction was a logical consequence of the time dilation caused by relative motion. We cannot accept one without the other.'

'Understanding different length seconds is bad enough.'

'That is because time does not operate independently of space and it only makes sense when you see how they work together. So here you are sitting in Spaceship Beetle in the stationary frame. You cannot measure that you are moving, but what about measuring the length of her cabin? You could use a tape measure or lay out rulers end-to-end, or you could take advantage of Maxwell's constant.'

'We could use light.'

A green beam of laser-light flashed in the darkness, crossing the Beetle from the middle of the rear window and illuminating a spot in the middle of the windscreen. The computer beeped '**6 light-nanoseconds/6 feet**' onto its screen.

'Did you really measure that?'

'Of course. Like the laser tape measure used by a builder, we are simply timing the journey of a light beam and converting it into distance using c. Now, what will you see if I come zooming past you at 90 percent of c measuring the inside of Spaceship Beetle II? When I fire my measuring-beam from my rear window, by the time it has crossed the six feet to where my windscreen should be –'

'It's not there any more.'

'My windscreen keeps moving forward through space as my measuring beam chases it down, and by the time my beam does catch up it will have travelled twice as far as your measuring-beam crossing Spaceship Beetle in your stationary frame.'

'You mean your beam ends up travelling through 12 feet of space, twice the length of the inside of your Beetle?'

'Yes. Which means it will take 12 nanoseconds to reach the windscreen and we should measure the inside of my Beetle to be 12 feet long.'

Danny smiled. 'But your time's running at half the rate of mine in the stationary frame so it will do it in six nanoseconds and –' He stopped smiling. 'Hold the horses, that can't be right.'

'That's right, that cannot be right. That means my cabin is three feet long.'

'Eh?'

'Light does not travel at two feet per nanosecond and cars do not double in length, and we would know we were moving if they did. Even though they should be taking 12 nanoseconds to travel 12 feet in our moving frames, our measuring-beams have to take six nanoseconds on our clocks to travel the six-foot length of the Beetle's cabin in our stationary frames. The only way that can happen is if space contracts in the direction of motion so that the distance our measuring-beams actually travel in our moving frames is halved from 12 to six feet.'

'You said your Beetle shrinks to three feet.'

'The only way that distance of 12 feet can become six feet is if the space inside the Beetle contracts to half its normal length in the direction of motion so that there is only three feet between the rear and front windows. Let's measure the length of the cabin again.'

The green laser flashed across the Beetle, beeping '**6 light-**

nanoseconds/6 feet' back onto the computer screen.

'According to the time dilation equation, in my moving frame in Spaceship Beetle II travelling at 161,000 mps or 90 percent of c, time is running at half the rate of yours in the stationary frame in Spaceship Beetle. For every nanosecond that passes on your clock, you see only half a nanosecond pass on my clock.'

'Because everything in your moving frame is travelling through twice the distance in space.'

'But the equation for speed is simply distance divided by time and the measured speed for light is always c, or one foot per nanosecond. So if, after six nanoseconds on your clock, you measure the time interval between the event of my beam leaving my rear window and the event of its arrival at my windscreen to be three nanoseconds on my clock, then my beam can only have travelled through three feet of space, not six. From your stationary frame Spaceship Beetle II's cabin is –'

'Three feet long,' said Danny, coughing up a laugh.

'If the time measurement in our speed equation gets smaller, then the distance measurement must get smaller by a corresponding amount so that every observer's measurements divide to produce c as the speed of light in all frames. My measuring-beam still travels twice the three-foot distance across Spaceship Beetle II in my moving frame, but it catches the windscreen after six feet instead of 12. The same applies to your moving frame in Spaceship Beetle, and in that way the speed of light is one foot per nanosecond in both our moving and stationary frames but our measurements of equivalent intervals of space and time are different.'

'Our beams travel six feet in six nanoseconds in our stationary frames and three feet in three nanoseconds in our moving frames.'

'The beams in our light clocks bounce straight up and down; we measure the speed of light to be one foot per nanosecond and the inside of our Beetles to be six feet long; and everything appears normal as though we are both in a stationary frame. And just as my time dilation equation shows how time intervals expand as you approach the speed of light, the Lorentz length-contraction equation shows how distances shrink by the same factor. At half the speed of light relative to an outside observer, my Beetle will be nearly 15

169

percent shorter in the direction of motion. At 90 percent of c she will be almost half her length. At 99.995 percent she will be 100 times shorter. And in the stationary frame of an atomic particle travelling at 99.99993 percent of c around the 16.8-mile tube of the Large Hadron Collider, it is the tube that is revolving and in that moving frame it is only 100 feet long. As we get closer and closer to the speed of light, the body contracts progressively until at c it disappears completely.'

'So from every different point of view or stationary frame, my time's running at a different rate and the Beetle's contracted by a different amount?'

'Each observer will come up with a different pair of measurements, but whatever the expansion of time there will be a corresponding contraction of space that cancels the effect and ensures they measure the speed of all light to be c and themselves to be in the stationary frame.'

'How can everything appear normal if the inside of my Beetle's only three feet long?'

'Everything looks normal because your eyeballs have contracted by half and refocused everything back to its normal length. And if you measure Spaceship Beetle's cabin with a 12-inch ruler, you will be measuring its three-foot length with a piece of wood that is only six inches long. Like time, space is not absolute and the distance between the marks on a ruler to the distance between planets is relative.'

'Does all this really happen?'

'It has to. If the effects of time dilation and length-contraction did not occur, our measurement for the speed of light would vary and we would know we are moving without reference to an outside body. It has to happen for us both to measure that we are at the centre of the sphere of light from your bulb in deep space. And in the continuing absence of any evidence for the Ether, it has to happen to explain the Michelson/Morley experiment. The fact is, if we did not adjust for both time dilation and length-contraction all sorts of modern equipment would not work correctly, from the GPS in your mother's car to your television set at home.'

Danny stared into the deep space of the Beetle's windscreen.

'It may sound crazy, but if we accept Maxwell's constant as a new law of physics then we have to accept that there is no absolute interval of time between any two events separated in time, and there is no absolute interval of space between any two events separated in space. And like atomic particles in an accelerator, you and I are living proof of it.'

That was Danny's problem; it didn't sound crazy enough any more.

'We should get back to planet Earth,' said Bert, with a gentle punch to Danny's shoulder. 'I have a Beetle to fix.'

*

Climbing out of the Beetle was like leaving a summer matinee, and Danny lowered the brim of his Stetson to block the slanting rays of the late afternoon sun. 'Bert?' he said, pulling up a chair. 'Can I go online?'

'Yes, of course. How did you get on with the generator?' Bert peered into the slot. 'Good job,' he said, clearing it off the table and turning the briefcase to face Danny. 'There may be some interference from the electromagnetic storms so do not worry if it is a little slow. I am going to check on the horses and fit this back.'

As Bert headed onto the plain, Danny activated the Google search engine and typed in *Albert Einstein*. There were over 140 million entries, but though the computer was slow connecting to some links, over the next hour site after site revealed picture after picture, and there was absolutely no doubt – it was Bert.

Scanning through the links he found confirmation of everything Bert had told him, from his job at the Swiss Patent Office where he formulated his Special theory of relativity in 1905, to his stance against Senator McCarthy's communist witch hunts in 1938, which put him under the scrutiny of the FBI until his death on April 18th 1955. He also learned of a troubled education that showed nothing of his future promise; the death threats and departure from Nazi Germany in 1932 for a life in America; his letter to Roosevelt in 1939 warning of the threat of a Nazi atomic bomb; and his declining the offer of the presidency of Israel in 1952.

At the bottom of one of the pages he found an obscure link titled

'*The Einstein Mystery*', which reported that conspiracy theorists had pointed to a number of inconsistencies in the account of Einstein's death. Apparently there was no official pronouncement as to the cause, other than that he had gone peacefully in his sleep after a period of illness, and a buried interview with one of his neighbours reported that he'd seen him drive off the week before, but he'd never returned and the car had never been found. What little family he had left had not been given the chance to view the body before burial because the authorities were busy removing its brain. And recently, some specialist had noted that this brain, which was kept at a small doctor's surgery in Wichita Kansas, appeared to be a few years younger than the body it supposedly came from and couldn't be his. There was even a grainy black-and-white photograph, supposedly of Einstein, taken in Moscow's Red Square in 1962, but Danny could tell it wasn't him. All this had led some theorists to argue that he had defected into anonymity in Russia, although most believed his outspoken political views had made him too many enemies and the FBI had '*rubbed him out*' – which in a sense they had. The official reply to the article from some spokesman at the FBI was as calculated as it was sarcastic: *Yes, and we didn't go to the Moon either.*'

Danny sat back in his chair with his eyes fixed to the screen, until a blurry image loomed behind it and they refocused onto Bert.

'Was there much interference?'

Danny's buried questions started rising, like zombies, and with their same dogged determination they grabbed Bert and pinned him to his chair. 'Where did you end up when you arrived in 2011?'

'Oh. Well I didn't know where I was, but it turned out to be a small Navajo Indian community called Cedar Ridge, northeast of here. I lived with them for almost a year and then bought the Missing Horse.'

'What did you buy it with?'

'It appears my death has greatly increased my worth. I found one of my old notebooks on relativity and $E = mc^2$ in the Beetle's glove box and sold it for a small fortune. After I had caught up with the advances in science, and upgraded the Beetle, I began trying to figure out how it could –'

'Does Morris know?'

'Nobody knows, except you.'

'Where are we now?'

'According to the Beetle's GPS, we are in a remote area on the South Rim of the Canyon about 40 miles from where we originally landed.'

Having been about to ask how he'd found him, Danny jumped to, 'Where did the horses come from?'

'The Missing Horse,' said Bert, as though Danny already knew all this. 'When I analysed the data on my briefcase computer and realised what had happened, I knew I would need transport to find you so I based myself at Rosie's and had Morris bring the horses there.'

Another question occurred to Danny, but he didn't really want to ask. 'How am I getting to the airport tomorrow?'

'I am going to drive you there. We can follow the trail left by the Sheriff.'

Never mind negotiating the bouncing path of the Sheriff's monster truck, Danny wondered how they were going to get the Beetle off its limestone platform. Staring out over the plain and losing himself in the distant hills, he didn't notice Bert lift himself out of his chair and bury himself in the Beetle's boot.

17

With its island-mountain summits blushing pink with the last kisses of sunlight, they set up the table and chairs on a flat rock at the edge of the Canyon. As they dined on thick steaks and fire-baked potatoes, Bert talked of his life on the world stage – driven not by the pursuit of riches and fame, but knowledge, world peace, and an unshakeable belief that one will lead us to the other.

Danny listened but with a question hovering at the back of his mind, like a gatecrasher at a party. It finally surfaced after the last of Morris' homemade apple pie had sunk and they sat watching the stars appear over the Canyon.

'Bert? I get the idea that the fixed speed of light means space and time are relative, but how does that explain what happened to us?'

'The speed of light means there can be no instantaneous interactions between bodies across space and it must take time to travel distance.'

'Because nothing can travel faster?'

'Nothing can ever reach the speed of light, let alone pass it.'

'Why not?'

'According to my time dilation and the Lorentz length-contraction equations, at the speed of light all mathematical analysis breaks down. We cannot measure anything because space has infinitely contracted and become one infinitely expanded interval of time. So you can accelerate from 99.99999 percent of c to 99.999995 percent, then on and on always creeping a little closer, but you can never reach 100 percent because you can never reach infinity and the speed of light is the maximum possible.'

'That's maths.'

'Okay then. We have seen that we cannot label an event with a universal now, which means –'

'I'm not sure about that either. I still don't see why we can't agree with an observer on the sun, or anywhere else, that the asteroid hit Jupiter at 7.40.'

'If we could send signals instantaneously to Jupiter and the sun, then we could all synchronise our clocks to the impact time of 7:40 and agree on a universal now. But that would be a simultaneous event for all observer's, and as we saw with our two strikes of lightening, that is not possible if they are in relative motion.'

'And everything's in relative motion.'

'In a moving frame a body travels an extra distance through space. That takes extra time because that distance cannot be covered at a speed greater than 186,000 mps. It is as though Nature has set a speed limit on the universe and nothing can accelerate to that speed, so instantaneous interaction across space is not possible.'

'Why can't something travel not instantaneously, but faster than light?'

'Forgetting the maths and the infinities at the speed of light, if a body could travel faster than light it would violate our understanding of causality. Remember when we were surfing on the beam of light showing your Town Hall clock striking midday?'

'We could only see that time because no later light beams could reach us.'

'What would happen if we could accelerate to move faster than that beam?'

'Would we catch up with the light beams that left before that one?'

'That's right. As we passed the wave fronts of earlier beams showing earlier times we would see the clock hands, and everything else, moving in reverse. In effect we would be traveling backwards in time, and to me that could not be right. Whether it is directly emitted or reflected, we use light to see events and we always see a penalty being scored after the ball has been kicked and not before. That places a limit on the speed at which any body can move and that limit is the speed of light because it regulates our possible observations of cause and effect.'

'How do you mean?'

'If you are separated from some event by an interval in space then you are also separated by an interval in time and no possible consequence of that event can have any effect on you before that time. Like the idea that the sun disappears right now.'

'We will keep receiving its light and heat for another 8.3 minutes.'

'In the same way, if you have been invited to a game of football on Jupiter and kick-off is in 30 minutes, even traveling at the speed of light, you are going to be five minutes late and you can do nothing to avoid it. You are separated in space, so you are separated in time, and anything and everything you do can have no consequence on Jupiter for the next 35 minutes.'

'What I see happen is what can happen?'

'In a sense, yes. Whatever you measure from your point of view, that is your reality because your observations and measurements are limited by the speed of light. That is the entire principle of relativity as it ensures that the laws of physics are the same for both of us, regardless of how we are moving.'

'What about the addition and subtraction of velocities?'

'I came up with a new equation that factors in the expansion of time and contraction of space to work out the relative speed between any two bodies and show that nothing can reach the speed of light. If I am in Spaceship Beetle traveling at 90 percent of c relative to you, and I launch Spaceship Beetle II away from me at 90 percent of c relative to me, and Spaceship Beetle II launches Spaceship Beetle III at 90 percent of c relative to her. From your stationary frame you will not measure Spaceship Beetle II to be moving at almost twice c and you will not measure Spaceship Beetle III to be moving at almost three times c. My new equation says you will measure them to be moving at 99.45 percent and 99.97 percent of c.'

'That's just weird.'

'Think what is happening from your point of view. When Spaceship Beetle flies past you at 90% of c, for every second that passes on your clock, you see only half a second pass on board Spaceship Beetle and in that half-second she travels only half the distance through space. When she fires out Spaceship Beetle II, for every second that passes for you, only one-hundredth of a second passes in her moving frame. You see her move forward a tiny distance of space and you measure her speed to be only a few thousand miles per second faster than Spaceship Beetle's.'

'Still below the speed of light.'

'When she fires out Spaceship Beetle III, you measure an incredibly squashed Beetle to be hardly moving any distance at all. No matter how many copy Beetles we launch or how many velocities we add together, from your stationary frame the expansion of time and contraction of space in each new moving frame will always result in you measuring a relative speed less than the speed of light.'

'Can we keep getting closer forever, even though we never reach it?'

'Theoretically, yes. But as you get closer, every tiny increase in the speed of your moving frame results in a huge increase in the expansion of time and contraction of space. Like your measuring-beam chasing after your windscreen, you would be on the edge of your seat at the next British Grand Prix if Lewis Hamilton were only seven seconds behind Nico Rosberg with seven laps to go and lapping a second quicker. If he were lapping only 0.00001 percent quicker, you might be tempted to leave before the rush.'

'It would take thousands of laps to catch up.'

'That is my Special theory of relativity,' said Bert, failing to suppress a quick bounce in his chair. 'It simply states that *the laws of physics are the same for all observers in uniform motion, and that time expands and space contracts by a proportional amount so that every constant velocity frame of reference measures the speed of light to be the universal maximum of 186,000 miles per second in a vacuum.*'

Danny pushed back his Stetson and looked up at the stars, with Jupiter shining brightly between them. 'So what happened to me?'

'You climbed into my Beetle down there,' said Bert, pointing his pipe into the black depths of the Canyon. 'You climbed out up here. That journey through space cost time.'

'Yeah, but not three days.'

'It does if you travel nearly 800 million miles via Jupiter at an appreciable fraction of the speed of light. And that is what you must have done because you have experienced the time dilation effects of that motion. In the 40 minutes that passed in the Beetle, three days, one hour and 17 minutes passed here on Earth in exact accord with the time dilation equation of my special theory for your relative speed.'

Bert followed Danny's eye line into the night sky and stood up. 'Come on,' he said, folding his chair and picking up the lantern. 'I think it is time we took a closer look at what happened to you.'

*

Waves of interfering light lined up on the briefcase screen to form a picture of Danny's nose and nostrils with his head narrowing into the distance, towards the Beetle's roof.

'Somehow you activated the JTFP, my Jupiter Test-Flight Program,' said Bert, hanging the lantern off the side of the Beetle's roof rack and joining Danny at the table in its pool of yellow light. 'At exactly 9:00 am on Tuesday July 19th, the JTFP found and measured the relative velocity of Star HD187462, which was moving away from the Earth at 161,000 mps, or 90 percent of c, in the opposite direction to a direct rendezvous with Jupiter. My Beetle started moving in the identical but opposite frame, activating a video and time link between the Beetle's main computer and my briefcase. This is your transmission from the fish-eye webcam in the Beetle back to the Earth.' Bert pointed to the top left corner of the screen. 'And this is my view of your time being recorded on your clock.'

On screen the Beetle suddenly filled with brilliant-white light, and as Danny's eyes slowly closed and his hand rose slowly to shield them, '**Beetle Time 09:00:00**' crept to 09:00:01 and crawled to 09:00:02.

Bert pointed to the middle of the black stripe across the bottom half-inch of the screen. 'This is Earth time being recorded by my briefcase computer.'

'**Earth Time**' was running at its usual rate, and as it flipped to 09:00:20, '**Beetle Time**' was just hitting 09:00:05.

Danny checked the speed-reading in the top right corner of the screen. 'If I'm doing 90 percent of c, my time should be running only half as slow as normal, I mean yours in the stationary frame. That's a quarter of the rate.'

'So what is going on? From my stationary frame on Earth, moving at 90 percent of c it will take you 40 minutes to travel the 390 million miles to Jupiter and your arrival time on my briefcase clock will be 9:40. But the time dilation equation of special relativity says only half that time will pass in your moving frame, so I will see an arrival time of 9:20 on the Beetle's clock. If we jump your transmission forward.'

Bert pressed a button on the briefcase console and the picture flashed forward in a blur. When it froze at '**Beetle Time 09:20:00**' little had changed in Danny's open-mouthed expression, except that it was now bathed in the deep rusty light of Jupiter's Great Red Spot.

In the black stripe at the bottom of the screen '**Earth Time**' had stopped at 10:15:00.

'Your clock should show an arrival time of 9:40,' said Danny. 'Why does it say 10:15?'

'Do not forget the Doppler effect. As you moved away from the Earth the electromagnetic light waves of your video transmission had to travel further and further to reach me.'

'Oh, they became stretched and their frequency went down, so you saw fewer of my events in every one of your seconds.'

'This slowing-down-of-time Doppler effect added to your actual time dilation such that, for every minute that passed for me only 16 seconds passed for you. Total the Doppler effect for your entire outward journey and instead of my clock recording the time of 9:40 at the event of your arrival at Jupiter, it recorded a time 35 minutes later –'

'Because that's how long it took for my transmission to get back to you.'

'Exactly. The electromagnetic waves showing 9:20 on your clock took 35 minutes to travel the 390 million miles back to me on Earth. Add that to the 40 minutes I recorded for your outward journey and the result is a total of 75 minutes passed in my stationary frame and I time your arrival to be at 10:15. For the duration of that outward journey your time ran at 27 percent or just over a quarter of that rate, and I see on your Beetle's clock that it took only 20 minutes in your moving frame. What happened from your point of view?'

'I didn't go anywhere.'

'From your stationary frame in the Beetle, the Earth went shooting away into space at 90 percent of c and Jupiter came shooting towards you at the same speed.'

'Then Earth time should be running slow compared to mine.'

'So if we look at my briefcase transmission from the Earth to the Beetle at the moment Jupiter arrived with you.'

Bert punched in another command, Danny disappeared, and the screen went blank save for the time readings, which remained but swapped places. In the black stripe at the bottom of the screen, '**Beetle Time**' advanced at its normal rate through four seconds then froze at 09:20:00, but '**Earth Time**' in the top left corner crawled through only one and froze at 09:05:33.

'Where's your picture?' said Danny.

'I did not return to camp and open my briefcase until 9:10 Earth time. But we can see that when you arrived at Jupiter at 9:20 on the Beetle's clock, it was only 9:05 on the Earth. From your point of view our situation was reversed and for every minute that passed in the stationary Beetle only 16 seconds passed in the Earth's moving frame. Total the Doppler effect at the event of Jupiter's arrival with you and the Earth's furthest point of recession, and instead of you seeing the time of 9:40 on my briefcase clock –'

'I see a time 35 minutes earlier because that was the last of your time signals to reach me.'

'That's right. My time of 9:05 was just arriving when you reached 9:20 on your clock in the Beetle. The wave showing my time to be 9:40 had only just left on its journey and was still 35 minutes away. For the duration of her 20 minute outward journey, Earth time ran at 27 percent or just over a quarter of the rate of Beetle time and you see on my briefcase clock that it took a little over five minutes in that moving frame.'

Danny pushed back his chair and dug two bottles out of the cool-box. 'How does any of that explain why three days have passed?'

'That was your outward trip. On reaching Jupiter the JTFP reversed direction and you started moving at 90 percent of c relative to Star BD+192777, which was moving in the exact opposite direction to a rendezvous with your original departure point on Earth – give or take a few miles. First of all, what did I see happen? From my stationary frame it will take you another 40 minutes to complete the return journey and I should record the time of 10:20 on my briefcase clock for your arrival back on Earth. Only half that time will pass in your moving frame and I should see you arrive back after 20 minutes of Beetle time, at 9:40 on your clock.'

'When I arrived at Jupiter it said 10:15 on your briefcase clock.'

'Which means I have to see your entire return journey take place in only five Earth minutes. That is exactly what happens because on the return trip we were moving towards each other.'

'The Doppler effect worked the other way round and you saw more events happen in my moving frame for every one of your seconds.'

'For every one of the 40 minutes that passed on my clock, I see only 30 seconds pass on your clock due to time dilation, but that 20-minute return journey is squashed into only five minutes of Earth time because of the Doppler effect.'

At Bert's command the screen jumped back to Danny's video, and as **'Earth Time'** in the black stripe climbed steadily at its normal rate from 10:15:01, in the top left corner **'Beetle Time'** raced ahead four times as fast.

Danny watched himself twitch and jerk on the screen. 'So for every minute of Earth time nearly five minutes of Beetle time passed for me.'

'And I measure your 20-minute return journey to occur in only five of my minutes so that you arrive back on Earth at 10:20 on my clock. What is your take on the return journey? In your stationary frame it said 9:20 on your Beetle's clock at the event of Jupiter's arrival with you, but what time was recorded on my clock at that event?'

'It was 9:05.'

'We know that my clock must read 10:20 at the event of Earth's return and yours 9:40, which means you have to see 75 minutes of Earth time squashed into 20 minutes of Beetle time.'

'I'll see your time run nearly four times as quickly as mine.'

Bert switched videos again, and as **'Beetle Time'** in the black stripe advanced normally from 09:20:01, and **'Earth Time'** raced ahead nearly four times as fast, Bert's face suddenly filled the screen, his worried blue eyes twitching and his jaw motoring up and down.

'From your point of view,' said Bert, 'the Earth will arrive back after a further 20 minutes and at 9:40 on your clock, but in that interval of time 75 minutes will pass on my clock and it will read 10:20. Ignoring the cancelled-out Doppler effect for the outward and return journeys, the rate of the passage of time was halved in the Beetle compared to the Earth exactly as predicted by the time dilation equation of special relativity for your relative speed.'

'Then only an extra 40 minutes should have passed on Earth?'

'Ah, yes. No, that's what the JTFP was supposed to do but it was over-ridden by the ERP, the Emergency Return Program, when you were still one minute away.'

'How did that happen?'

'You must have held down the red key.'

'You told me to!'

'I said *do not hold it down*. I said there would be big trouble if you hold down the red key.'

'You didn't say it, you squeaked it and the sound and picture kept breaking up.'

'Those blasted solar flares disrupted the transmission.'

'I thought you said, if there's trouble hold down the red key. I didn't know what was happening. I just wanted to get out of the car.'

'I know,' said Bert, putting his hand over Danny's. 'I am sorry. None of this is your fault. The problem is, the ERP searches for the fastest body it can find moving in the opposite direction needed to get back to roughly the same point on Earth and out of spaceship mode as soon as possible. Moving at 90 percent of c, you traveled the first 381 million miles of the return journey in 19 minutes in your moving frame and 38 minutes in Earth's stationary frame. But you traveled the last nine million miles relative to a cosmic particle moving at 99.9999973 percent of c and the time dilation was such that, in that last minute on your clock an extra 4,358 minutes passed on mine. As a result a total of 73 hours and 17 minutes passed on Earth for your 40 minutes in the Beetle.'

'That's not much of an escape from an emergency!'

'It is if there is a fire and you have to get out of the car as quickly as possible. But what else can we see must have happened on your journey? If you had completed the 390 million-mile trip to Jupiter in 20 minutes, you would have traveled at 325,000 mps, nearly twice the speed of light. You did not, of course, because in your moving frame you did not travel through 390 million miles of space.'

'You mean the distance to Jupiter shrank by half as well?'

'Special relativity says the space between the events of your leaving and returning to the Earth contracted by half in the direction of motion. On your outward journey, from my stationary frame you traveled the correct distance of 390 million miles in the correct time of 40 minutes for your set speed of 90 percent of c. In your moving frame you traveled the correct distance of 195 million miles in the correct time of 20 minutes for that set speed. The same length-contraction occurred during the first 19 minutes of your return journey, plus a greater contraction during your superfast final minute but we need not to go into the maths.'

Danny stared at Bert's saggy face frozen on the screen. 'Why aren't we squashed up in our videos? Is it because the cameras were always looking head on?'

'Even if we had been filmed from the side, we would still look normal on our videos because the thickness of our webcam lenses had contracted by half in our moving frames. To observe each other's relative contraction we would need powerful telescopes so that we could directly see each other from our stationary frames.'

'Why didn't you send me a message on the outward trip?'

'It was 9:05 on my briefcase clock when your return journey started and I did not return to camp until 9:10. If you had received my 9:12 transmission before 9:20 on your clock then my electromagnetic waves would have had to travel faster than the speed of light and you would have received the message before I sat down to film it.'

'So what was meant to happen to me on your expedition?'

'Nothing would have happened to you. We would have analysed my theory to find the answer to this mystery, and then to find out what an outside observer would see, at the end *I* was going to do the JTFP to demonstrate the 40-minute time dilation to you. Not the other way round with over three days dilation.'

With his last, desperate pockets of resistance blown away by a barrage of evidence and logic, Danny didn't know what else to say.

He didn't know what to say as they closed down the briefcase, unpacked the sleeping bags and smothered the fire. He didn't know what to say as they lay beneath the canopy of stars, galaxies and nebulae, with the passing dots of satellites gliding in between. It was only as he felt himself on the verge of sleep that he finally spoke.

'Bert,' he said quietly.

'Uh?' grunted Bert, pulling himself out of his own leap in time.

'I just wanted you to know, I'm glad I'm here.'

'So am I, Danny ... So am I.'

18

Danny woke up to the warmth of the sun, the smell of sizzling bacon, and the strange fact that it was Saturday 30th, his last day at the Grand Canyon, and he couldn't wait to tell Tom all about it.

Over breakfast, it grew into a list of just about everyone he knew. 'Then there's my drama teacher, Christina. She'll probably add a couple of gruesome murders and make it our next play. And our neighbour, Mr Rahman, subscribes to *Bizarre* magazine.'

'You haven't mentioned your mother,' said Bert, dissecting his bacon into rectangles.

Danny laughed. 'She'll never believe me.'

'Do you think anyone will? I have not even told Morris.'

'Tom will.' Danny eye's narrowed then popped wide. 'There's the Beetle's video of the trip to Jupiter.'

'That was either looking at you or the Beetle's roof.'

'Oh yeah. Mr Rahman grew up on a farm and told me saw a chicken get beamed into an alien spaceship, so he might.' Danny's head dropped back to his plate. 'Moving without moving and Jupiter coming here does sound even more unbelievable.'

'That depends. If all motion is relative, what does moving actually mean?'

'It means everything's moving, but everything isn't supposed to suddenly take off,' said Danny, waving his fork and launching a piece of bacon off to the side.

'Or does relativity mean that everything must be standing still relative to everything else.'

Danny stopped searching the ground. 'Are you serious? I thought the whole point of relativity was that there's no such thing as standing still.'

'The principle of relativity holds that all movement is relative and there is no absolute uniform motion.'

'Apart from the speed of light.'

'Why apart from the speed of light? Its fixed speed is the basis of relativity so the principle should apply to light as well. And if every observer in every moving frame measures the speed of light to be c, then from the stationary frame of a beam of light everything else must be moving at c.'

'But everything's moving at different speeds relative to everything else.'

'Through space alone, yes, but everything is moving through space and time. It could be the other way round, but in the same way that positive kinetic energy and negative gravitational energy combine to the total energy of the system, so positive time and negative length vary in direct proportion always adding up to the combined maximum of the speed of light.'

'As time-intervals grow, space shrinks by the same amount.'

'Which means they are completely interchangeable. We can treat time as another dimension of space and view our world as existing in four dimensions.'

'The spacetime continuum.'

'The problem is, our three-dimensional brains cannot visualise a fourth dimension.' Bert swung his leg out from under the table and drew the wooden ruler from his boot. 'We can understand one dimension,' he said, drawing a single line in the sand. 'We can add two more dimensions with two more lines, each at a 90-degree angle to the first.' He drew another line to form a right angle and then jammed the ruler, upright, in the corner. 'Where would a fourth line fit at 90 degrees to each of the other three?'

'It can't.'

'We cannot visualise a step up to four dimensions, but what about a step down? How long is this ruler?'

'Twelve inches.'

'How long is it in two dimensions?'

Pinched between his forefinger and thumb, Bert held the ruler square to the rays of the sun. As he pointed to its full-length shadow on the side of the Beetle and rotated it slowly, Danny watched the shadow get shorter and shorter until it almost disappeared.

'A shadow is a two-dimensional projection of a three-dimensional object. The ruler has a fixed three-dimensional length, but that length can vary from 12 inches to nothing when seen from different angles in two dimensions. In the same way, every body has a fixed four-dimensional *spacetime length* where 186,000 miles of space is equivalent to one second of time, and each observer slices up their three-dimensional view of that four-dimensional length into different intervals of space and time.'

'So our world's like a shadow of four-dimensional spacetime?'

'That's right. In a stationary frame, like inside Spaceship Beetle on your trip to Jupiter, you measure all your motion to be through the time dimension of spacetime and it passes at its normal rate. But in a moving frame events are not separated only by time.'

'There's an extra interval in space.'

'That motion through the spatial dimensions uses up some of your motion through the time dimension. As the speed or rate of your passage through space increases and intervals of space contract, intervals of time expand by a proportional amount and the rate of your passage through time decreases.'

'Until I reach the speed of light and space contracts into an infinite expanse of time.'

'Except that nothing can ever reach that speed, so every interval of space involves some interval of time and every interval of time involves some interval of space, but every body is always moving at the speed of light through four-dimensional spacetime. That is why we live in a spacetime continuum; that is why every observer measures all light to travel at c, and that is why all motion is relative.'

As Bert's last words drifted through Danny's brain, a thought struggled to form. He stared at the western horizon trying to coax it out then suddenly abandoned it completely. 'Indians!' he cried, jumping out of his chair.

On their side of the Canyon, in direct line with Danny's fork, an inverted cone of smoke rings expanded high into the sky.

'Oh, yes,' said Bert. 'It must be important. I hope it has nothing to do with us.'

'What do you mean?'

Bert checked the plain, causing Danny to do the same. 'This is

Indian land and we are not supposed to be here,' he said, looking back to the smoke rings. 'After breakfast we should head west and see if we can spot anything.'

'In the Beetle?'

'I think it would be best on horseback. But we should pack everything away so that we are ready to leave for the airport on our return.'

Bert's eyes rolled to meet Danny's, but flicked away as soon as they did.

*

They rode for many miles, mapping the Canyon edge like Spanish Conquistadors, and galloping across peninsulas like outlaws on the run. It was only as the ground began to disappear that they realised they'd been fooled by a bend in the Canyon walls, and after dismounting to watch the smoke rings repeat their simple pattern – above the north side of the Canyon and not their own – they climbed onto the horses and started heading back.

'I feel like a bounty hunter who opted for the *Dead* option,' said Bert, moving Marilyn further from the edge and leading Charlie by his reins beside her. 'You cannot be more comfortable like that.'

Danny's bottom and inner thighs had started protesting as soon as he'd sat in Charlie's saddle back at the Beetle. By the end of the outward journey they were chained to the gates of every thought waving placards and shouting, '*STOP!*'

'This is great,' said Danny, raising his cheek from Charlie's thigh, his upper body and arms hanging upside-down on one flank and his legs dangling down on the other.

He let his head fall back to Charlie's soft chestnut hair, and as his body rocked gently across his rump, he remembered that something had been niggling at his brain then jack-knifed as he worked out what it was. 'Hold the horses!' He tried to twist his head up, but all he could see was Bert's upside-down boots. 'If all motion's relative and we can say the Earth and Jupiter were moving, why didn't you miss out on three days instead of me?'

'I wondered how long it would take you. Yes, it is called the

Twins Paradox because, even though they both experience time dilation effects in their own moving frames, identically aged twins in relative motion cannot each age less than the other.'

'What's the answer?'

'My special theory is called that because it is restricted to a special form of motion: uniform motion in a straight line. But what happened when we drove down that track after I picked you up on your first day?'

'I could tell I was moving because we were accelerating.'

'You could tell without reference to an outside body, which means acceleration is a form of absolute motion.'

'I thought there was no absolute motion.'

'There is no absolute uniform motion because the laws of physics are not affected by constant velocity motion and every observer can say they are in the stationary frame. But if you play table tennis or perform experiments in an accelerating frame, the results will vary greatly depending on how your velocity is changing. Not only do you know that you are moving –'

'I know how I'm moving.'

'And your accelerated motion is absolute. But there is a maximum speed in the universe and you cannot go instantaneously from a stationary frame on Earth to a moving frame relative to the Earth. It takes time to travel distance and your velocity has to change over time.'

'You have to accelerate away.'

'And you have to change direction or come to a stop and reverse to get back. That is why the Twins Paradox does not apply to my special theory. On your outward journey to Jupiter we each measured the time dilation effect on the other's passage of time, and as long as we keep moving away from each other in that same relative frame, we will never be side-by-side again and we will never be able to make a direct comparison of our clocks. We will always be separated in space, which means –'

'We will always be separated in time.'

'But when you accelerated away from the Earth, decelerated to a stop at Jupiter and accelerated back again, you actively changed your frame of reference relative to the Earth. Those accelerations distinguished your moving frame from my stationary frame because your motion was no longer relative it was absolute.'

'I didn't feel any acceleration of any kind,' said Danny, hauling his body up Charlie's side and swivelling his legs to straddle his rump. He unhooked his Stetson from the pommel and sat up, looking like one of Bert's cherry lollipops. 'So why was it me?'

'Because you were down there, you are now up here, and over three days have passed on Earth for your 40 minutes in the Beetle. No the problem is, my special theory does not include accelerated motion, and that raises another problem. What happens to the missing energy?'

'What missing energy?' said Danny, crossing his forearms onto Charlie's neck and hooking his feet above his tail.

'Newton's second law says a body will accelerate by the same amount for every unit of force it receives. Special relativity says, even if we keep pumping out the full energy of Spaceship Beetle's rocket engines, we cannot apply a continuous force to accelerate her to the speed of light.'

'Isn't it like when she fires out Spaceship Beetle II and she then fires out Spaceship Beetle III and they don't have time to get very far?'

'The force acts for a shorter time interval in the moving frame, so each Beetle experiences less and less acceleration as it approaches c. But if she is not accelerating by the same amount for each unit of force she receives, where does the unused energy from the rocket engines go?'

'It has to be conserved.'

'The law of conservation of energy says it has to be transferred into something else, but what? The answer is surprisingly simple: $E = mc^2$.'

Danny smiled, until it occurred to him. 'We're not got to do more maths, are we?'

'No, you have all the information you need to understand that *energy is equal to the mass of a body multiplied by the speed of light, multiplied by the speed of light again.* We know that energy and momentum are fundamentally connected through the laws of conservation, and in 1772 a French chemist called Antoine Lavoisier used a sealed flask and a very precise weighing machine to show that mass is conserved as well. Whether it was rotting fruit, burning paper or rusting metal, he discovered that any loss of solid mass was balanced by the release of gases into the flask and the overall mass and weight remained unchanged. Mass is clearly connected to momentum and energy, but what exactly is mass?'

'It's the amount of stuff in a body.'

'How do we measure it?'

'It's the same as its weight, so we can weigh it.'

'That's right. Newton's law of gravitation says that gravity is proportional to mass, so we can measure any unknown mass by seeing how many known masses, like a kilogram, are needed to balance a set of scales so that gravity acts equally on both. We call this measurement of mass using gravity, *gravitational mass*. What if you are standing on the surface of the Moon or Jupiter?'

'My mass is the same but my weight's different on each because the strength of their surface gravity is different.'

'But if we cannot weigh you to measure your gravitational mass in different gravitational fields, how else can we measure your mass? Why does everything fall at the same rate?'

'If you double the mass of a body, you double the force of gravity, but you also double its inertia and halve the effect of the force.'

'So a body's mass determines both its weight and its inertia, which means we can use comparisons of inertia to measure mass in the same way we use comparisons of weight on a set of scales. If my Beetle's engine can accelerate her two tonnes to 40 mph in four seconds, and yet the same engine force applied for the same time can only accelerate another body to 20 mph, then we know that its mass is four tonnes. We call this measurement of a body's mass using inertia, its *inertial mass*.'

'Is it the same result whichever way you do it?'

'Yes. Inertial and gravitational mass are the same because mass is directly proportional to both. Returning to Spaceship Beetle trying to accelerate to the speed of light. My new equation for the addition of velocities tells us that the same force applied for the same interval of time has less effect in the moving frame as you approach c. So if the same size force is having less effect on a body then it must be because –'

'It's got more inertia.'

'And a body's inertia or resistance to a force is a measure of its inertial mass, which means –'

'It's got more inertial mass.'

'And inertial and gravitational mass are equivalent, which means the energy going into a body increases its weight, which in turn means –'

'It gains mass.'

'That's right. Spaceship Beetle's missing energy is turned into extra mass.'

'Will she get bigger?'

'No, she will get smaller in the direction of motion due to Lorentz/Fitzgerald length-contraction as her atoms are squeezed tighter together. The point is, if an input of energy increases a body's mass then that energy must itself have weight and be equivalent to a certain amount of mass. Conversely, any mass must be equivalent to a certain amount of energy, which means, like space and time, one can turn into the other. The speed of light is the conversion factor between space and time, and the speed of light squared is the conversion factor between mass and energy. A body in the stationary frame will always have its own *rest mass*, but the body's *relative mass* will vary depending on its speed relative to each observer. My second is not your second, my inch is not your inch and …'

'My kilogram is not your kilogram.'

'By factoring in the time dilation and length-contraction effects of special relativity, I came up with a new equation which shows that the relative mass of a body in a moving frame increases at the same rate as intervals of time.'

'That means I weighed 100 kg on the trip to Jupiter?'

'But you could not feel or measure the increase in your stationary frame because the mass of your bones and everything around you had increased by the same proportion. In a moving frame at a relative speed of 99.995 percent of c your mass would be 100 times greater than your rest mass. And a proton circling in the LHC at 99.99995 percent of c is over 1000 times more massive than identical protons in the stationary frame of the laboratory. As you get closer to the speed of light, for every tiny increase in your speed there is a huge increase in your mass, requiring a huge increase in energy to accelerate you further.'

'Oh-h. The same force produces less acceleration because it's moving a heavier and heavier body.'

'Exactly. At the speed of light itself, your mass would become infinite, it would take an infinite amount of energy to get you there and so –'

'Nothing can accelerate to the speed of light, ha!' Danny smiled. 'Why is it the speed of light squared?'

'If you remember, Gottfried Leibniz discovered that $E = mv^2$.'

'Yeah, and a car's breaking distance is four times as long if you double the speed. But why?'

'What is the maximum speed any body could achieve?'

'When it's moving at the speed of light.'

'So a body's maximum kinetic energy would be when it is moving at c. But the body's rest mass is also equivalent to energy, so the total energy of the body is the product of its maximum possible kinetic energy *and* the energy locked up in its rest mass if that were accelerated to c.'

'Oh right, so you have to multiply its mass by the speed of light again.'

'That's it,' grinned Bert. 'Mass is solid energy and one can be converted into the other using $E = mc^2$. And because the speed of light squared is such a huge, huge number, there is an enormous amount of energy tied-up in every tiny mass.'

'Like an atom bomb.'

The light slowly faded from Bert's eyes, and as he turned away his face bore only wrinkles of time. Crickets chirped above the soft beat of horses' hooves then were suddenly drowned out by them, as Marilyn moved through the briefest of trots and started cantering across the plain.

Ignoring his bottom, Danny pulled himself into Charlie's saddle and they soon matched Marilyn's gait, as she galloped across good ground and cantered across bad, but always a few hundred yards behind. For a while he imagined he was the Rango Kid in search of his next adventure, but it was a very short while. This was the real adventure and he, Danny Robinson, was living it. He'd made the right choice of camp; the choice his dad would have made. It was hard to accept that it was coming to an end.

The horses came together at the Beetle, parked on its rock outcrop like an off-road exhibit at a car show, and they removed their tack and started brushing them down.

'I am sorry,' said Bert, the plastic brush in his hand making tracks along Marilyn's white flank. 'I made one mistake in my life when I signed that letter to President Roosevelt advocating that the atomic bomb should be built. Had I known that the Nazis would not succeed in producing one, I never would have lifted a finger.'

Danny was busy working on Charlie.

'$E = mc^2$ pointed the way to the atomic bomb, but it is the same energy that millions rely on across the world. It explains the mystery of radioactivity, and how stars like our sun produce huge amounts of heat and light for billions of years. It confirms why the bouncing electromagnetic energy beam in your light clock has momentum, just like a ping pong ball, and why nothing can reach the speed of light.'

'It still doesn't explain my trip to Jupiter,' said Danny, standing back to admire the *DR* he'd branded with his brush onto Charlie's thigh.

'No, but it is linked to another problem that does.' Bert gave the horses a slap, and they skipped out onto the plain to graze. 'My special theory does not include accelerated motion, but our universe is dominated by acceleration because of gravity.'

'Gravity causes everything to accelerate towards everything else.'

'Which means gravity is a form of absolute motion not explained by the theory. But we know that Newton's universal law of gravitation is wrong and that –'

'How do we now know it's wrong?'

'Because nothing can travel faster than the speed of light. Not even a helicopter.'

Danny was about to query the suitability of the example when the sound hit his ears, his hat blew off, and a Bell Executive, four-seat, twin-turbine rose out of the Canyon behind the Beetle. It buzzed and twitched as it studied them though bulging Plexiglas eyes, like a giant dragonfly contemplating lunch, then looping into the Canyon and back over the plain, it landed on an area of sand 30 yards from the Beetle.

'Who's this?' said Danny, as the engines whined down and friction started working on the blades.

They did not find out until the rotor had come to a complete stop and it was immediately clear why. A crown of brilliant-white, black-tipped eagle feathers emerged majestically, eight feet above the ground and two feet above the sharp brown features down the sides of which it flowed. Dressed from head to foot in tasselled buckskin decorated with zigzagging patterns of coloured beads and a porcupine-quill breastplate, the Indian walked towards them and stopped a few feet short.

'How,' he said, slowly raising his hand and swatting at a bothersome fly, 'the hell did you guys get out here?'

19

'I'm Marvin,' said the Indian, extending his hand to Bert.

He was in his early twenties, with a powerful frame and an angular face dominated by a swooping beak of a nose and chiselled chin, but softened by deep brown eyes that appeared to be permanently smiling.

'Or I should say Moon Rise. It's this powwow,' he confided, sounding every bit the modern student. 'My tribe, the Hualapai, are hosting a gathering this weekend to celebrate our religion and culture. The Chief's very strict, and being his grandson he gives me an especially hard time.'

'Hello Marvin. I'm Bert and this is Danny.'

'Hi,' said Marvin, shaking Danny's hand.

'Is that your helicopter?' said Danny.

'No, but today I get to use it.'

'We apologise for coming onto your land uninvited,' said Bert. 'I am afraid we experienced a few difficulties and ended up here by accident.'

'You were spotted by our scouts, and Grandfather was wondering if you are the NASA scientists from the newspaper.'

'That's right.' Bert gave Danny a nudge with his elbow.

'Yes,' said Danny, shifting his weight casually onto one leg, 'we are they – them – I mean –' He smiled at Marvin and said the next thing to come into his head. 'What are those smoke rings saying?'

'Welcome and peace,' said Marvin, looking to the signal then at each of them in turn. 'Welcome and peace. My grandfather, Chief Shooting Star, invites you to join us.'

'We would consider it an honour,' said Bert.

'Grandfather was hoping to meet you. A fiery comet marked his birth and he has spent his life listening to the stars. I have an errand to run but I'll be back to pick you up in about an hour, just after midday.'

'We're going in the helicopter?' said Danny.

'If you don't ride so often, it's the only way to get around these parts.' Marvin looked past them to the Beetle. 'How *did* you get out here?' He watched Bert and Danny glance at each other, open their mouths, then close them. 'Top secret eh? I'd better be going.'

With a final dazzling smile Marvin loped back to the helicopter; its engines whirred into life; its blades into a shimmering disc; and it lifted off and peeled away to the north.

'This could work out perfectly,' said Bert.

Danny turned his attention to the plain and the search for his Stetson. 'What do you mean?'

'If we play our cards right, we may be able to get you a lift to the airport.' Bert raised his hand to his eyes for one last look at the helicopter. 'I am not sure we would get my Beetle off that rock.'

*

After checking the roof rack and securing the ratchet straps, Danny had planned to sit on the limestone wall at the edge of the Canyon and wait for the return of the helicopter, but Bert had other ideas. 'Jump in,' he said, climbing into the Beetle, 'and let's get to the bottom of your journey.'

As Danny closed his door, the screens rose up the windows, leaving only the pinpoint twinkles of the star field to slowly penetrate his eyes.

'You have experienced the time dilation of a flight to Jupiter,' said Bert, his outline emerging with the stars, 'but you did not experience the accelerations necessary to distinguish your moving frame from the Earth's stationary frame so that you alone experienced that dilation. Why not? The answer cannot lie with either Newton's laws or my special theory –'

'Yeah, why is Newton's law of gravity wrong?'

'According to Newton, the gravitational force between two bodies depends on their mass and the distance between them, but it is *instantaneous* action-at-a-distance and not affected by time. So if the sun disappeared now, the Earth would immediately start moving in a straight line, like Io when you let go of the string.'

'But nothing can travel faster than the speed of light.'

'Which is why the Earth would not abandon its orbit for another 8.3 minutes and Newton's law cannot be right according to my special theory, even though that theory does not include an explanation of the effects of acceleration and gravity.'

'Hold the horses!' said Danny, snapping round to face Bert. 'Shouldn't I have been weightless on the trip to Jupiter?'

'Exactly. Which brings us to the question: what is gravity?'

'It's the attractive force between bodies.'

'But *why* does a body's mass pull everything towards it? In 1905 I did not know the answer, but I believed that no form of motion should be absolute; there should be no special frame of reference from any point of view; and the laws of physics should remain the same for all observers, even if they are accelerating. Then I had the happiest thought of my life.' Bert paused to do an impression of the wide-eyed, open-mouthed grin he'd adopted at the time. 'I realised that gravity and acceleration are equivalent and we cannot distinguish between them.'

'What about now? I can feel my weight and gravity, but I'm definitely not accelerating because I think I'm standing still.'

'Look at it this way.' Bert tapped the keypad, replacing the starlight with the roof light. 'Let's imagine you are floating weightlessly inside Spaceship Beetle with window screens raised so that you cannot see outside. How are you moving?'

'I must be doing a constant velocity deep in empty space. Or I'm in orbit.'

'Or you could be falling into the Grand Canyon. For that brief period before you caught up with the Beetle, you were accelerating under the influence of gravity alone and you could not feel your weight. What was my pipe doing?'

'It was spinning in mid-air.'

'If we ignore the spinning, all the objects in your accelerating frame remained in the same place relative to you.'

'Because we were falling at the same rate.'

'But if you cannot feel any forces acting on you and everything around you stays in the same position relative to you, then you are entitled to say …'

'I'm standing still.'

'So you have no idea if you are traveling at a constant velocity in outer space or accelerating towards the ground in the gravitational field of the Earth because the effects in each frame are identical. What would you say is happening if you suddenly sink back into your seat and feel your full 50 kg weight again?'

'It'd be like when the parachutes opened and we stopped accelerating. Or I could just be sitting on the ground.'

'What if you lower the screens and find that there is no planet Earth or anything else, just empty space?'

'Then I've nothing to measure my motion against.'

'But you should be weightless and you are not. Even though it cannot be the result of gravity, you are feeling a force of 500 newtons between you and your seat.'

'Oh, I could be accelerating and feeling the effect of my inertia trying to stay at the same speed.'

'That's right. When thrusters fire on the bottom of Spaceship Beetle and she starts accelerating in the direction of her roof, you will remain floating where you are and the surface of your seat will rise up and accelerate your body in that direction.'

'I'll get progressively heavier, like when we fell into the Canyon.'

'If Spaceship Beetle were constantly accelerating up at g, you would feel your normal weight in your seat as though you were sitting on the surface of the Earth. What happens when you hold up my pipe and the bowling ball and release them at the same time?'

'The Beetle's floor accelerates up to meet them at the same rate as gravity.'

'And from your stationary frame inside Spaceship Beetle they appear to fall side-by-side to the floor because of gravity. They remain pinned to the floor by their apparent weight in the same way that you are pinned to your seat by yours, and you cannot measure that you are accelerating without referring to a body outside.'

'Does that mean gravity and accelerated motion are not absolute because I can still say I'm in the stationary frame?'

'Exactly! In the stationary frame of a falling apple, the Earth accelerates up to meet it, the tree accelerates away in the opposite direction and the principle of relativity applies in both the stationary and moving frames. That is my *Principle of the Equivalence between Gravity and Acceleration*, which simply states that *we can swap any gravitational field for an equivalent rate of acceleration that will cancel its effect so that we think we are standing still in empty space, and we can swap any acceleration in empty space for standing still in an equivalent gravitational field.* In other words, you could play a game of table tennis in a sports hall accelerating at g in deep space as though you were standing on the surface of the Earth. But a game in a free-falling sports hall would be identical to trying to play weightlessly when moving at a constant velocity in deep space.

'It would be even harder if the sports hall was spinning like your pipe.'

No, quite the opposite. Even if a body is rotating at a constant speed, it's direction and therefore its angular velocity are still changing over time, which means it is still accelerating and that is still equivalent to gravity. Let's say you are in Spaceship Beetle with the screens raised and you are being pressed into your seat with a force of 500 newtons. Where are you and how are you moving?'

'I could be sitting on the Earth or accelerating straight up at g in space.'

'Would it surprise you if you lowered the screens and discovered that Spaceship Beetle is sitting on the sidewall of a large centrifuge, like Mr Boggle's Sticky Wall, rotating in deep space?'

'Oh, I'll feel the effect of my inertia trying to remain in a straight line.'

'If the centrifuge is accelerating round at the constant rate g, Spaceship Beetle will be pinned by her inertia to the wall; you will be pinned to your seat; and when you release them, my pipe and the bowling ball will fall at g in a straight line to the floor.'

'The Beetle's still going round on the wall so won't they fall at an angle and hit the door? And like Io and Ganymede, they'll just move off at whatever speed they were doing before I let go.'

'To someone outside, observing your moving frame, they do move away at a tangent to their orbit on the centrifuge and at a constant speed. But in your stationary frame inside Spaceship Beetle with the screens raised there is no rotation. The pipe and bowling ball were standing still in your hands so they must be accelerating relative to you. And they do not head towards the door because the Beetle continues to rotate in the same direction as they fall, and their resultant path is to fall vertically from your hands to the floor, as though you are sitting on planet Earth. That is why scientists talk of using a centrifuge to create artificial gravity on long space flights.'

'Like the revolving spaceship in *2001 A Space Odyssey*.'

'We measure this artificial-gravity effect whenever a body is accelerating; especially with rapid changes in direction, like jet pilots performing acrobatics in their planes.

'They feel heavier because they pull more g.'

'The point is, we could play a game of table tennis in a sports hall if it were stuck to the wall of a centrifuge rotating in deep space as if we were playing on the surface of the Earth. And if the sports hall were falling in the Earth's gravitational field and spinning like my pipe, we can in theory introduce a suitable strength gravitational field that will cancel the extra inertial effects of that rotation.'

'So the force I feel when I'm accelerating at g is not gravity but my inertia?'

'They produce the same resultant force because your inertial and gravitational mass are the same and your body responds equally to each force or effect.'

'I thought a body's inertia was its natural motion, not a force?'

'Ah, ha! Now we are getting somewhere. If inertial effects are the same as gravitational forces then maybe gravity is also natural motion.'

'You mean gravity isn't a force either?'

'There lies our final hurdle and the answer to our problem. What causes gravity and why do we feel its effect as a force when in the presence of mass? Ha! But now we can work out the answer. If gravity is not a force –'

Danny and Bert jumped out of their seats as a loud bang reverberated from the Beetle's roof. Bert pressed a key and the window screens cleared to reveal Marvin without his feathered war bonnet.

'Hey?' he smiled. 'Are you guys coming or what?'

Three minutes later, with the helicopter buzzing beneath its invisible spinning wings and a knot of fear – that nervous companion to every thrill – tightening in his stomach, Danny secured himself in the seat behind the pilot and grinned at Bert beside him.

Marvin pointed to the chunky headphones hanging from the roof, and Danny settled a pair over his ears, muting the engines and roar of the wind to a background flutter, as gentle as the vibrations passing from the airframe to his body and forgotten as the headphones crackled into life.

'Hi there,' said the pilot. 'Everyone hear me okay? My name's Earl. I'll just wind this baby up to full power and we'll be on our way.'

Danny nudged Bert with his knee. 'It's Earl,' he mouthed, revving the throttle on the handlebars of an imaginary motorbike.

The helicopter lifted off, dropped its nose and charged towards the edge of the Canyon, the ground filling the bubble screen and rushing beneath them in a blur.

'It can be mighty disconcerting when your altitude suddenly goes from a few feet to thousands,' drawled Earl through the headphones. 'You might want to close your eyes round about now … that's what I do.'

Earl didn't mention that the second part of his strategy for dealing with the void was to follow it straight down, and Danny groaned as his stomach got left behind, then grinned as his downward acceleration cancelled the effect of gravity and whatever parts of him that could floated freely around the cabin.

He felt a tap on his arm and Bert grinned back, his frizz of white hair standing up either side of his headphones and his limbs hanging on invisible strings.

The helicopter powered into the earth, passing the top layer of white limestone, hundreds of feet of sandstone in streaks of red, yellow and grey, and thousands of feet of purple limestones and pink shales – before Earl pulled them out of the dive and everything collapsed with a weight greater than gravity.

Continuing their descent through winding canyons and sloping valleys, they swept over lifeless bulges of smooth lava, then plunged over the sharp edge of the innermost canyon and down its vertical wall. The mighty Colorado River swirled beneath them, thick with debris from a thousand crumbling walls, and following an invisible track in the air, they started banking around its sweeping bends.

Then they started to climb.

Up the jagged walls of the inner canyon. Up barren dunes of rock washed smooth by countless rains and scored with the valleys of their progress to the river. Up huge cliffs separated by giant scree-slopes, rising like an old flight of stairs that have never been swept, and up the valley between soaring tabletop mountains.

Rounding an Aztec pyramid crowned with a crumbling temple dome, the Canyon's north wall appeared, like a lorry heading for a gnat. Courses of red-stone bricks and blocks flashed past the bubble screen as Earl powered the helicopter up its face, and they climbed and climbed until the earth coughed them out over woodland of towering pines.

The helicopter levelled out, skimming feet above the treetops and 150 above the forest floor visible between them. Expanding circles of thick smoke billowed high into the sky straight ahead, and they burst over a final fringe of pine and found their source.

In the middle of a grassy plain the large signal-fire smoked heavily with four Indians throwing a blanket on and off. Dozens of teepees ranged around it, like snow-covered volcanoes erupting black sticks, and everywhere the bright plumage of Indians. Adults cooking, eating, dancing and tending horses in makeshift corrals; and children running and waving up to them as they raced to the landing-zone west of the camp.

Earl air-braked the helicopter to a stop, 70 feet above the ground but close enough to the crowd of upturned faces to make them gasp in the wind from its blades. As the last child arrived, he powered it into the airborne-equivalent of a doughnut, and spinning like a seedpod from a sycamore tree, they floated down and landed.

Danny looked out at the brown, smiling faces, their features joining them in race as clearly as their dress separated them by tribe. Plain leather loincloths jostled next to tasselled buckskin leggings decorated with beads and porcupine quills, and cotton shirts and skirts trimmed with coloured zigzags and stripes. But they all wore necklaces and bracelets on their bodies, and whether skewered through buns, tied to shiny-black plaits, sprouting from moccasins or hanging from bows, they all carried at least one white and black-tipped eagle feather.

Taking his cue from Marvin, Danny removed his headphones and put on his Stetson. 'We're the only cowboys here,' he muttered to Bert, forgetting Earl, the only real cowboy in the front.

Marvin turned round as he adjusted his war bonnet, and the tips of its feathers folded along the glass. 'This powwow is a celebration of all Native Americans and our culture. We are here to worship our Mother, the Earth, in ancient rituals of private song and dance. But don't worry,' he smiled at Danny. 'You are my brothers and we welcome you as honoured guests … we won't scalp you till later.'

The helicopter had been the centre of the children's attention throughout its landing and wind-down to silence, but as two cowboys emerged, it moved with them in a boisterous giggle – with most of the giggling directed at the young cowboy who was walking strangely and kept fidgeting with his hat.

'Grandfather sends his apologies for being unable to greet you,' said Marvin, guiding Bert between the teepees. 'This is a rare opportunity for the elders to talk of the problems facing our people. But tribal council meetings can be long affairs so he has asked that you enjoy our celebrations and he will meet with you later this afternoon.'

'It's like a film set,' Danny whispered over Bert's shoulder, as the milling splendour of the camp opened before their eyes.

'Really? I've never seen one.'

'No-o. I mean how it's so real yet not real. Or not anymore.'

'Reality is seldom what it seems.'

In case Danny needed a reminder that Bert was right, it came in the form of a tug on his chaps and a little Indian boy, no more than Alice's age, looking up and announcing, 'It's Saturday', as though he'd just learnt how to say it, then running back to the safety of the crowd.

Marvin said something in the deep sound bites of his native tongue, and as the children dispersed into the camp, he turned to Danny. 'I told them you were a great white spirit and that if they did not show some respect, you would come in the night and haunt them in their dreams.'

'Did you?'

'No.' Marvin smiled. 'I told them lunch was ready.'

'Is it?'

20

The blue-green water lapped around Danny's knees as he washed his face and upper body in the pool, then settled around his calves, as flat and clear as glass.

Ferns clung to the creek's limestone walls, marking the course of intermittent waterfalls down their sides, and sprinkled with scarlet flowers, they hung from their crevasses and ledges, like the Gardens of Babylon. Dense moss cushioned the rocks around the banks of the pool, and as the broad leaves of an elder tree scattered the sunlight across its surface in a million twinkling stars, Danny wriggled his toes and sank deeper into its sandy bottom.

He wasn't sure how they'd found it. After watching four ritual dances with Bert and Marvin, he'd slipped away and been picked up by a gang of Indian boys for games of *Stickball*, a sort of field hockey, and *Shinny*, which was a bit like lacrosse. It was late afternoon when he spied Bert marching across the plain in search of him, and leaping out of the long grass in his underpants – finger-smears of mud across his face and chest and an eagle feather lodged in the belt around his head – he fired the straw arrow from his bow to join the hail heading in from all directions. After the old man had died theatrically and chased his whooping brother warriors away, they stumbled upon the creek and its forgotten pool.

'We have reached the final hurdle,' said Bert, from a nest of long grass at the water's edge.

'To what?' said Danny, wading deeper into the pool.

'To finding the explanation for your journey to Jupiter. We have established that gravity is equivalent to a simple acceleration and that inertial and gravitational effects are the same. So when an apple falls from a tree, or you let go of my pipe and the bowling ball, rather than viewing their equal rate of fall as gravity producing an equal resultant force on each, we can say it is due to the relative effect of the ground accelerating up to meet them as they remain standing still next to each other in space.'

Danny's Y-fronts halted his advance. 'I get that if I'd been accelerating at g I would feel my full weight and think I'm still on Earth, but the Beetle's speed was a constant 161,000 mps. And she didn't travel there and back roof first.'

'But what if gravitational acceleration is a body's natural motion through space, like inertia.'

'How can it be natural when inertia wants to keep everything moving the way it is, not start accelerating?'

'Let's imagine you are weightless in Spaceship Beetle in orbit around the Earth. At every instant the force of gravity accelerating you towards the ground is balanced by your inertia trying to make you move in a straight line at that instant.'

'One cancels out the other and there's no resultant force.'

'But if you are in orbit around a planet yet *think* you are moving in a straight line in deep space, then the two paths must be equivalent.'

'What do you mean equivalent?'

'I mean the curved path of a body in orbit is actually a straight line.'

Danny laughed. 'A straight line can't be curved because then it wouldn't be straight.'

'What is a straight line?'

'Isn't it the shortest distance between two points?'

'What happens to parallel lines?'

'They remain parallel and never meet.'

'What do the three angles of a triangle add up to?'

'180 degrees.'

'Those are the basic rules of two-dimensional or *plane geometry* worked out by a Greek mathematician called Euclid in 300 BC. They are based on light traveling in straight lines, but light does not always follow the shortest path between two places. Point at a stone on the bed of the pool.'

'There's one there,' said Danny, pointing through the water surface a few feet in front of him.

'Is that where the stone actually sits?'

Danny looked back to the stone but he didn't need to. 'It's in the wrong place.'

'When light enters a different density medium at an angle, it is bent at the junction by a property called *refraction*. Passing from water to air, the light waves are reflecting off the stone at a steeper angle through the water than required for the direct straight-line path to your eyes, and on reaching the surface they are refracting to a new, shallower angle to complete the journey. To you, the direction of the source of the light is down the straight-line of that second, shallower angle and the stone appears to be further away and less deep than it actually is.'

'Doesn't light travel more slowly in water?' said Danny, reaching down for the stone and finding it closer than it looked and his fringe dipping unexpectedly into the water.

'Maxwell's equations confirmed that light slows down as it moves through denser media, but a French mathematical genius called Pierre de Fermat had explained why this caused refraction way back in 1650. He came up with the *principle of least time,* which says that light will always take the quickest path between two points, even if that is not the shortest path. This has been nicknamed the Baywatch Principle, apparently after some television program.'

'It's about lifeguards. Girls, mostly.'

'That explains it. If you are a lifeguard and you are quite a distance down the beach when you spot me drowning, you are not going to jump into the sea and start swimming along the direct straight-line path.'

'I'll run along the beach first.'

'You can run a lot faster than you can swim, so without thinking about it you would run down the beach at one angle and then enter the water to swim the rest of the way at a second angle to reach me in the minimum possible time. The light reflecting off that stone is doing the same as your beach rescue but in reverse. It takes a steeper angle to the surface to minimise its time in the water, then refracts to a new angle to maximise its time in the air and reach your eyes in the shortest possible time.'

'How does it know to do that?'

'Fermat's principle of least time is also called the principle of least action because Nature does not like energy to be used unnecessarily. You sit down whenever you get the chance. Atomic particles always settle to their ground state, which is when they are vibrating with the least use of energy. And light always travels the quickest path to minimise the energy of its self-creating process. So a straight line is not the shortest distance between two points, it is the quickest, and because nothing can travel faster, light must always travel in straight-line paths even when it is moving in a curve.'

'You've lost me on this one,' said Danny, turning his attention to the pond-skaters performing around his thighs. 'How can a straight-line path be curved?'

'What is the quickest path you can fly between London and Phoenix?'

'Ah,' twigged Danny. 'That's not straight because we have to follow the curve of the Earth.'

'We can connect any two places on the Earth's surface by drawing a great circle around its circumference. The shorter portion of that circle is the shortest distance between those two places and the resulting curve is called a geodesic, which is the path your flight from London to Phoenix followed to consume the minimum amount of fuel.'

'If we built a tunnel through the Earth that would be quicker because that is the real straight line and the shortest distance.'

'Let's drop down a dimension. If we are two-dimensional beings living on the surface of a large three-dimensional sphere, then our world appears to be a flat two-dimensional plane. To us there is no up and down, and we cannot tunnel through anything, so how could we measure that we are living on a sphere?'

'If we keep traveling in a straight line, we'll end-up back where we started.'

'Using Euclid's laws we can work it out without circumnavigating the globe. Let's say we each head directly north from two places, *A* and *B*, both situated on the equator, which to us is simply the straight line running east to west across the middle of our plane world. We set off, keep heading north in a straight line and to our surprise we meet each other at *C*.'

'The North Pole?'

'That is simply another place on the flat surface of our world, so how on Plane did we meet? Somehow our parallel paths form a triangle connecting the three places, and yet its base angles at A and B already add up to 180 degrees.'

'The top angle at the North Pole makes it more.'

'Which means Euclid's laws of plane geometry do not apply and our world cannot be flat. Even though we are only two-dimensional beings, we could work out that we must live on a two-dimensional surface that somehow curves back on itself in a higher dimension and has some strange three-dimensional shape that we cannot begin to imagine. As three-dimensional beings we understand that this two-dimensional surface curves to form a three-dimensional sphere and our straight-line paths are in fact geodesics, which is why we meet up at C. But the point is, the behaviour of a straight line on a two-dimensional surface reveals the geometry of that surface or how it is shaped.'

'Okay, for two-dimensional beings a geodesic is the shortest path.'

'If we jump back up to three dimensions, let's imagine you are floating around Spaceship Beetle as she moves through empty space at a constant velocity in the direction of her roof. What will happen when you measure the length of the inside of Spaceship Beetle?'

'The beam will hit the front screen after six nanoseconds.'

'It crosses the Beetle in a straight line, parallel to the floor, and hits the front screen after six nanoseconds, even though the front screen continues to move up at a constant velocity as your measuring-beam travels through space from the rear screen?'

'Yeah. Like the beam in my light clock, my measuring-beam's sharing the Beetle's motion and its momentum keeps it the same height above the floor.'

'And it hits the middle of the front screen, directly opposite the source on the rear screen, at position X. Now suppose Spaceship Beetle starts accelerating at g in the direction of her roof and you sink into your seat and feel your 50 kg weight. When you fire off your measuring-beam again, its momentum will keep it moving up at the Beetle's velocity at the moment it leaves the rear screen, but it will *not* share the acceleration that is still occurring to the Beetle.'

'Then it won't travel parallel to the floor and it'll hit the front screen below position X.'

'From your stationary frame, the beam curves down as it crosses the Beetle and hits the front screen lower down, at position Y.'

'Doesn't that go against the principle of relativity because I'd know I was moving without looking outside.'

'Not if you measured the same thing to happen when you are sitting in the Beetle in the stationary frame on planet Earth. And because acceleration and gravity are equivalent, that is what you have to measure in order for the principle of relativity to be upheld. If light curves in an accelerating frame then light must also curve in a gravitational field.'

'You mean my measuring-beam bends towards the floor because it's being attracted to the Earth's surface by gravity?'

'Light is electromagnetic energy and $E = mc^2$, which means light must fall into a curve in the gravitational field of planet Earth at the same rate as a bullet fired from a gun or the International Space Station in orbit.'

'So a beam of light doesn't always travel in straight lines.'

'The path of a beam of light is how we define a straight line. It is the quickest possible path between any two points in space.'

'But we're talking about our three dimensions, and if light follows a curved path then it's not the shortest distance. Like tunnelling through the Earth to get from London to Phoenix, if the light followed that shorter straight-line path, it would be quicker.'

'Then why doesn't your measuring-beam take that path as it crosses the Beetle in an accelerating frame or in a gravitational field?' said Bert, jumping to his feet at the edge of the pool. 'There is nothing to stop it and no need for a tunnel.'

'Yeah, why doesn't it go in a straight line?'

'It is moving in a straight line. The path of your measuring-beam must be the shortest distance across the three-dimensional space inside my Beetle because if there were a shorter and therefore quicker route, Fermat's principle of least action says your measuring-beam would have taken it. There cannot be a shorter path because nothing can travel faster than light, which means nothing can travel between those two points along a path that is less curved and shorter than your beam and your tunnel cannot exist.'

Danny waded into the dappled shade of the elder tree. 'It still doesn't make sense.'

'It does not make sense in our obvious view of space, but the behaviour of light has shown us that we are wearing another blindfold. If a beam of light follows a curved path through three-dimensional space then the straight-line, shortest path through that space must be curved, which means the *space itself must be curved*. In the same way that straight lines can tell our two-dimensional beings that the surface of their world is curved, we can use the straight-line behaviour of light to give us a map of the geometry of three-dimensional space.'

'Isn't my measuring-beam curving because it's being attracted towards the Earth by gravity and not because space might be curved?'

'Flip that around and we have the answer to one of the greatest mysteries in science. If the geometry of space is curved by the presence of masses like planets, then the force we call gravity could simply be the effect of everything accelerating naturally as it falls down the valleys and contours of that curved space.'

'You mean gravity's like rolling down a hill?'

'Exactly!' cried Bert, launching his arms to the sky and his body into the water. 'Mass does not create an invisible, long-distance force called gravity. Mass causes space to curve or *warp* around it and your acceleration is the natural motion you experience as you fall down the curved geometry of that three-dimensionally warped space.'

'So there really is a warp factor nine, like in *Star Trek*?' said Danny, rising onto his toes as Bert waded towards him.

'Not in use as an engine, but yes. Space is bent and folded, or warped, by the presence of mass and energy.'

Danny brushed his hands through a flotilla of elder leaves. 'I'm still not sure what you mean by curved space. There's no surface, is there, so what are we falling down?'

'We can get a better understanding if we look at it in two dimensions.' Bert rested his hand on Danny's shoulder and started sucking a leg out of its boot. 'What happens if you put two bowling balls at either end of a rubber sheet stretched out on a frame, like the bed of a trampoline?'

'They'll roll into the middle.'

'Were they attracted towards each other by some mysterious force acting at a distance? No, it was simply due to the geometry of the surface of the rubber sheet caused by their mass. If we remove one, the remaining ball will sit in a depression, or well, in the surface of the rubber sheet, and the sloping sides of the well will be steeper nearer the ball then curve up at a shallower angle as you move away. What will happen to the path of a marble if you roll it past the bowling ball?'

'Its direction will be changed by the depression.'

'If it is moving quickly its path will curve only slightly, but if it is rolling at just the right speed for its position in the well, it could roll all the way around the bowling ball.'

'Like an orbit.'

'If we could remove the force of friction, the marble would keep rolling around the wall of the well in a fixed orbit with a velocity that depends only on its initial speed and distance from the ball.'

'The same as gravity.'

Bert emptied the left boot and threw it onto the bank. 'Jumping back up to three dimensions,' he said, placing his other hand onto Danny's shoulder and reaching down for his right, 'a body's mass creates a three-dimensional *gravity well* in the actual fabric of space. The body's gravitational field can be seen as the sloping surface of the well, with the strength of gravity or acceleration at each position within the field depending on the curvature of space at that position. Out deep in space away from any mass, like the outer edges of the rubber sheet, space is flat. Your measuring-beam travels directly across the Beetle and you move in a straight line with the natural motion of your inertia. Around a massive body, like planet Earth, space is curved. Your inertia is accelerated at g into the natural attraction we call gravity and your measuring-beam accelerates down at the same rate as it crosses the Beetle.'

'Hold the horses,' said Danny, jumping round and almost spilling Bert into the water. 'Light can't accelerate! If my measuring-beam follows a curved path because it's falling at g then it's got to travel further than six feet and go faster than c to still cross the Beetle in six nanoseconds?'

'There can be only one explanation,' said Bert, throwing the other boot. 'Special relativity has fused space and time into four-dimensional spacetime, so if the rate of your motion through space is changing then the rate of your motion through time must be changing as well.'

'Time's slowing down again?'

'Time must slow down in an accelerating frame to allow your measuring-beam to cover the increased distance caused by the acceleration so that you think you are in the stationary frame. That is the only way the energy of light can be subject to gravity in the same way as mass and not violate Maxwell's constant or the principle of relativity.'

'Is it the same time dilation as in the special theory?'

'The principle of relativity applies to both uniform and accelerating frames so any form of relative motion will dilate time. But the extra motion in an accelerating frame produces an additional *gravitational time dilation* dependent not on the body's motion but its position.'

'Why its position?'

'Let's look at a simple accelerating system. Imagine you are rotating in the middle of the raised floor of Mr Boggle's Sticky Wall and you start crawling towards the outside wall. As the distance you must travel with each rotation increases, your circular speed or angular velocity increases, which means your acceleration increases and you feel a greater centrifugal effect trying to pull you towards the wall. In other words, your rate of acceleration depends on *where* you are on the centrifuge and that rate will determine the extra time dilation you experience at that position.'

'Why's that a gravitational time dilation?'

'As you crawl to the outside wall, if you lay rulers end-to-end and measure the radius of the centrifuge, and on reaching the wall you measure its circumference in the same way, what will you discover? In Euclidean plane geometry the ratio of any circle's circumference to its radius is 2 x pi written as 2π. But according to my special theory, your rulers contracted across their width as you measured the radius, but as you measured the circumference –'

'Oh, they contracted along their length.'

'In your accelerating frame the circumference is short compared to the radius and you are measuring a warped circle that curves up, like the bottom of a gravity well. Again, the accelerating frame of the centrifuge mimics a gravitational field with the extra degree of gravitational time dilation, or *time warp*, depending on the degree of space warp or equivalent acceleration at that position.'

'Time warps are real as well?'

'Time flows at different rates for different observers in a gravitational field, even if they are not in relative motion. And because the strength of gravity or equivalent rate of acceleration increases as you approach the centre of mass, whether it is a bouncing light beam, a wristwatch or the vibrations of an atom, all clocks run more slowly on the surface of the Earth than they do out in space.'

Bert threw his arms wide, causing Danny to flinch at the prospect of another hug.

'That is my *General theory of relativity*. Published in 1916, it is an extension of the special theory to include accelerating frames and gravity, and says simply that *the presence of energy and mass warps the fabric of four-dimensional spacetime to produce the natural acceleration we call gravity and a proportional slowing down or warping of time so that light travels one foot in one nanosecond in ALL moving frames.*'

Bert grinned at Danny, but his mouth remained closed and there was something different about his pale blue eyes. 'We have reached a completely new understanding of our universe and we have done it all with the power of thought.'

'Is that it? The final hurdle?'

'That's it.'

Danny showered Bert with droplets as his hands flew out of the water. 'It still doesn't explain my trip!'

'Ha! It explains everything!' cried Bert, scooping two handfuls right into Danny's face.

Danny was going to voice his whole-hearted disagreement but settled for a water fight.

He'd been dunked twice by the old man and was hanging onto his bony leg, trying to up end him, when the ululating voice of a young Indian boy rang out from the top of the creek. Turning immediately to wade out of the pool, Bert shouted back a reply.

'Since when can you speak Indian?'

'This is my home,' said Bert, picking up his boots. 'Come on. The Chief is waiting. I will go ahead and change and meet you outside his teepee.'

21

'With this sacred pipe you will walk upon the Earth. The bowl of the pipe is of red stone; it is the Earth. The stem of the pipe is of wood and this represents all that grows on the Earth. These four ribbons hanging here on the stem are the four quarters of the Universe.'

As Chief Shooting Star returned to unhurried silence – with Marvin's translation a beat behind – Danny sat on the earthen floor of the teepee trying to decide if he was older than Bert.

The Chief's large, flat nose had carried its structure far better through time, and firm skin stretched over cheekbones as proud as mountains. But his almond-shaped eyes didn't shine quite as bright; his face was creased as though it had been ironed against the ground rather than dragged there; and having lost the support of his teeth, his mouth had adopted a permanent frown.

'This tobacco comes from the whites. We mix it with the bark from the Indian trees and burn it together. So may our hearts and the hearts of the white men go out together and be made good and right.'

Danny had expected him to look more splendid than Marvin, but a single black-tipped eagle feather crowned thin, white hair as it parted along the middle of his head and weaved into plaits either side. His feet were bare, and simple buckskin leggings and shirtsleeves poked out from the faded blanket falling around his huddled frame.

'It is the pipe that connects us with the Great Spirit. As we smoke the pipe we should remember the importance of having a sacred centre within us and that this sacred presence is represented by the pipe.'

The Chief lit the three-foot, finely carved pipe with an ember from the fire and presented it to Bert, sitting beside him in a pair of white cotton trousers. After Bert had taken his time over two deep draws, Danny lined up the pipe and the bowl glowed as he sucked in the smoke.

Feeling certain that his eyes shouldn't water and his cheeks weren't supposed to inflate, he tried to look natural, but a thick mushroom cloud burst from his mouth and headed for the smoke hole, funnelled by the sapling trunks interlocked above his head. Managing to hold onto the smoke a little longer at the second attempt – and release it with a little more control – he passed the pipe to Marvin with the relief that at least he hadn't coughed.

Marvin started coughing and didn't stop until he'd handed the pipe to his grandfather and the Chief had smoked it dry. He spoke to his grandson as he emptied the contents into the fire.

'Grandfather says I have been away too long,' said Marvin. 'He also asked me to explain the talking feather. There is an old Indian saying: *listen or your tongue will keep you deaf.* As children we were taught that a pause giving time for thought is the truly courteous way of beginning and conducting a conversation and this is embodied in Grandfather's eagle feather. Whoever holds it may speak and it is passed around in turn to all who wish to speak. In this way we have acquired the habit of carefully arranging our thoughts.'

With the talking feather held lightly in his hand, Chief Shooting Star's mouth straightened into a smile. 'Friends and brothers it is the will of the Great Spirit that we should meet together this day. He has taken his garment from before the sun and caused it to shine with brightness upon us. The Great Spirit is within all things, the trees, the grasses, the rivers, the mountains, and all the four-legged animals, the winged peoples and the peoples of the stars. This is the Indian theory of existence but there are many different ways of seeing the world. Tell me of the white man's scientific view and what you have learned from the Universe. You must speak straight so that the words may go as sunlight to our hearts.'

The Chief passed the talking feather to Marvin; Marvin passed it onto Danny; and Danny, wanting to pass it straight on to Bert, clutched the feather, like a microphone, and swallowed drily. 'I'd … er … like to thank Chief Shooting Star for inviting us here today.'

Marvin spoke quietly at his grandfather's ear and the old man nodded.

'So ... the universe. What have we, that is scientists, like us —' Danny's eyes beamed out an SOS to Bert as his finger darted back and forth between them.

Bert glanced to the talking feather and sent tiny nods of encouragement back.

'Or better still, science. What has science learned of our universe.'

As Danny surveyed the expectant smiles of his audience, like a nervous best man at a wedding reception, it occurred to him that Marvin hadn't translated any of that. A nanosecond later it occurred to him why and deciding he'd better start doing this the Indian way, he closed his mouth and started to think.

'Things are not always as they appear. Sitting here now we think we're standing still, but everything in the universe is moving through space and we can only measure how something's moving relative to each point of view. This natural motion means you don't stand still and only move when you receive a force; you keep moving at a constant speed in a straight line until you receive a force, and that makes you accelerate in the direction of the force at a proportional rate.'

Danny stopped for another think.

'So when two bodies collide we can't say who hit who because each gets hit by the other and receives the same force in the opposite direction. And gravity must be the same and result from both masses, but it depends on the distance between the bodies as well. Oh yeah, and it's also proportional to your inertia, which is why everything falls at the same rate.'

That was Newton covered. Now what ...

'Then we measured the speed of light and discovered it's the same from every point of view. That's like watching Marvin ride towards you at full gallop, but when you get on your horse and ride towards him, or chase after him when he goes past, he approaches at the same speed as before.'

Which means ...

'Oo! The distance the light travels and the time it takes must be different for everyone so that they all measure the same speed and think they're standing still. And that means there's no fixed amount of time between any two events and no fixed length or distance between them either.'

Danny beamed around the teepee, until he remembered he was supposed to know all this. 'Yes, so …' He stopped for another think.

'Both space and time are relative to how a body's moving because they are joined together into four-dimensional spacetime. We can't picture it, but it's like time's another direction of space we can't see, and you're moving at the speed of light but your motion's split into proportional amounts of each. You could go on a trip to Jupiter traveling at 90 percent of c, and even though it only takes 40 minutes on your clock, over three days would have passed on Earth.

'How can that happen – you ask. As you move faster, time intervals become longer and space shrinks by the same amount so you don't notice anything. But why didn't you feel any accelerations? When you accelerate at g your inertia produces the same affect as gravity, and you can be falling in orbit and think you're standing still out in space because inertia and gravity cancel out, so gravity isn't really a force.' He glanced at Bert. 'According to Einstein's general theory of relativity, it's just natural acceleration as everything falls in the curved spacetime around a planet or whatever.'

How did he get to that?

'Oh, space is curved because energy and mass are equivalent as well, so light curves towards the Earth, like a bullet. And because nothing can travel faster, that curved path is the shortest distance through space and space itself must be curved.'

Danny paused for another smile of surprise, but unfortunately, not for a think.

'But how does that explain your trip to Jupiter? The journey was at a constant 161,000 miles per second straight ahead, and if we were the ones who experienced the time dilation then we must have changed our frame relative to the Earth to avoid the Twins Paradox. So how does the curved space and warped time of general relativity explain how I managed to accelerate up to that speed, come to a stop and accelerate back the other way when I felt nothing but my constant weight in my seat?'

Danny was unaware of the bemused look on Marvin's face and the vacant expression on the Chief's, but as he stared at the proud smile on Bert's, he realised he'd directed the last question straight at him and it had sounded a lot less rhetorical than the others. 'Yes, um,' he coughed, turning back to the Chief, 'these are the ... er ... questions facing scientists, like us,' he started with the finger again, 'and science ... today.' He thrust the talking feather at Bert.

'Let me tell you of the life of a star,' said Bert, a familiar sparkle glinting in his eyes and the talking feather tickling lightly at his toes. 'A star forms when gravity squeezes together a huge globule of hydrogen gas until its core explodes into nuclear reaction, fusing pairs of hydrogen atoms to form helium and releasing enormous energy into space as waves of heat and light.

'Every star performs a balancing act between the force of gravity pushing in and the force of its nuclear explosions pushing out, and its fate depends on its size. When the star's hydrogen fuel runs out, the helium core collapses until helium atoms start fusing to form carbon. Nuclear reactions outside the core balloon the outer shell and a stable *Red Giant* is formed. When the helium runs out, the core collapses again, blowing the outer layers of the red giant into space. The carbon core is left as a *White Dwarf* star: a ball the size of the Earth with a teaspoon of its mass weighing more than a tonne.

'Our sun will eventually collapse into a white dwarf, but we have never found one with more than 1.4 times the mass of the sun as they all collapse further, fusing carbon atoms into heavier elements like oxygen and iron. When these nuclear processes end, the core collapses, blowing away its outer layers in a huge explosion called a supernova. Left behind is a core fused entirely into neutrons: a *Neutron star*, only a few miles across, with a surface gravity of over a billion g and time running at a fifth the rate on Earth. But we have never found a neutron star with a mass of more than three suns, yet there are billions of stars over 100 times more massive than our sun.

'Einstein's gravitational field equations show that a star with over three solar masses will continue to collapse until it reaches a critical radius called its event horizon. The enormous gravitational force at that distance from the centre of mass warps time infinitely and wraps space around the collapsed star, trapping its light in orbit and devouring any matter that strays near. It has become physically cut off from our universe in space and time. It has become a *Black Hole*. With masses up to the equivalent of 18 billion suns, these island universes offer a tantalising glimpse back to the birth of our own.

'In 1924 an American astronomer called Edwin Hubble used a new technique to measure the distance to the fuzzy patch of light in the night sky called Andromeda and discovered that it was hundreds of thousands of light years further away than the stars in the Milky Way. Hubble had proved the existence of galaxies beyond our own, but his measurements revealed that a Doppler effect was stretching the wavelengths of light from every galaxy, shifting their colour towards the red, lower-frequency end of the visible spectrum. There could be only one explanation; the galaxies are all moving away from the Milky Way and each other, like spots on an inflating balloon.

'There lies the origin of the Big Bang theory of the universe. If all the galaxies are moving away from each other then we can run the clock back to a time when they were all together in a small region of space. Like a collapsing star, this huge concentration of mass would be squeezed at almost the speed of light into an infinitely dense atom or quantum of pure energy. Thirteen point eight billion years ago that quantum of energy exploded in the Big Bang.

'As my colleague mentioned, Einstein has shown us that our universe exists in four-dimensional spacetime. Space is not the static backdrop to motion but a dynamic and elastic part of it, and we have to abandon the idea of time as a flowing river and think in terms of *block time*. We cannot say what events are in the past; we can only say what events have passed for us. For others those events still lie in their future, just as there are events that lie in our future that for others have long passed. The past is out there. The future is out there. All time is out there in spacetime, and like any other physical quantity it can have a beginning and an end.

'The Big Bang was not like a bomb. It was not an explosion of some *mass* at some *time*, some *where* in space; it was pure energy *creating* time, space and matter as it exploded. This tightly warped ball of spacetime has been expanding ever since, separating the mass into separating galaxies, and with their light waves stretching to longer wavelengths not through a Doppler effect but as the nothingness of space stretches in all directions faster than the speed of light.

'We have known for hundreds of years that our sun is not the only star in our universe, and for over ninety that the Milky Way is not the only galaxy. Technological advances in astronomy are now confirming that there are planets around almost every star, and Earth-like planets in the habitable orbital zone around many nearby stars. We have opened a new chapter in cosmology and life in our universe ... what is out there, waiting to be found?'

Bert winked at Danny and passed the feather to the Chief.

The Chief pulled his toes out of the sand and tickled them with the feather, as Bert had done. 'It is good for the skin to touch the Earth and its life-giving forces. It gives me a feeling of being close to a mothering power. I will tell in my way how the Indian sees things. The white man has more words to tell you how, but it does not require many words to speak the truth.

'I am one of Nature's children. I was born in her wide domain. The trees were all that sheltered my infant limbs, the blue heavens all that covered me. When I was ten years old I looked at the land and the rivers, the sky above and the animals around me and could not fail to realise they were made by some great power.

'The Great Spirit is in all things. He is our Father and his spirit pervades all creation, but the Earth is our Mother and every part of her is sacred to my people. She shall be my glory; her features her robes and the wreath about her brow, the seasons, her stately oaks and the evergreen, her hair, ringlets over the earth – all contribute to my enduring love of her. Wherever I see her, emotions of pleasure roll in my breast and swell and burst, like waves on the shores of the oceans, in prayer and praise to the Great Spirit who has placed me in her hand.

'The Indian thinks with his heart. He is of the soil. He once grew as naturally as the wild sunflowers; he belonged just as the buffalo belonged. The white man thinks with his head; his heart away from nature has become hard. There is no quiet place in his cities, no place to hear the leaves of spring or the rustle of insect wings. The Great Spirit told us, you red people will see the secrets of Nature. The day will come when you need to share the secrets with other people of the Earth. The time to start sharing is today.

'I hear what the ground says. The land waits for those who can discern their rhythms, each continent, each river valley, the rugged mountains, the placid lakes, all call for relief from the constant burden of exploitation. Everywhere the white man has touched the spirit of the Earth it is sore.'

The Chief dropped his head and for a long time said nothing.

'As a young man I had a dream. I was standing on the highest mountain of them all and round about beneath me was the whole hoop of the world. And I saw that the sacred hoop of my people was one of the many hoops that made one circle, wide as daylight and as starlight. I saw that anywhere is the centre of the world.

'Sometimes dreams are wiser than waking. On the hoop of life there is a place for every species, every race, every tree and every plant. With all things and in all things we are relatives. But being born as humans is a very sacred trust because of the special gift we have, which is beyond the fine gifts of the other living things on Earth. We are able to take care of them. Humankind have not woven the web of life, we are but one thread within it. We are the physical mirroring of the total Universe upon this Earth, our mother, and whatever we do to the web, we do to ourselves. It is this completeness of life that must be respected in order to bring about the health of this planet.

'But the ones that matter most are the children. We do not inherit the Earth from our ancestors, we borrow it from our children, and when we walk upon her, we always plant our feet carefully because we know the faces of our future generations are looking up at us from beneath the ground. We never forget them.'

The Chief reached over the fire and placed his hand over Danny's.

'My hand is not the colour of your hand, but the Great Spirit made us both. We are in this together my friends, the rich, the poor, the red, white, black, brown and yellow. We share responsibility for Mother Earth and those who live and breathe upon her, but if you have one hundred people who live together, and if each one cares for the rest, there is one mind.

'There has been too much talk. Your words circle like soaring birds which never land. I will try to catch them and take them back for my people to hear. As we are going to part we will come and take you by the hand and hope the Great Spirit will protect you on your journeys and return you safely to your friends. But first I will sing this song to help you understand the Indian picture of all creation. Songs

224

are thoughts sung out with the breath when people are moved by great forces and ordinary speech no longer suffices.'

With his grating voice rhythmically rising and falling, the old Indian started to sing.

> *'With beauty before me, I walk*
> *With beauty behind me, I walk*
> *With beauty below me, I walk*
> *With beauty above me, I walk*
> *With beauty all around me I walk*
> *It is finished in beauty*
> *It is finished in beauty*

'The song is very short because we understand so much.

'Shadows are long and dark before me. I shall soon lie down to rise no more. But what is life? It is the flash of a firefly in the night. It is a breath of a buffalo in the winter time. It is as the little shadow that runs across the grass and loses itself in the sunset. I shall vanish but the land over which I now roam shall remain. Oh that I could make that of my people and of my country as great as the conceptions of my mind when I think of the Spirit that rules the Universe. You look at me and you see only an ugly old man, but within I am filled with great beauty. I sit on a mountaintop and look into the future. I see all peoples living together as one.'

The old Chief passed the talking feather back to his grandson and the deepest silence fell around them, as though the entire camp had been frozen and only they drifted on through time.

'Does anyone wish to speak again?' said Marvin.

Nobody reached for the feather.

The flap of the teepee rustled, and the young boy's head popped through, his eyes wide and his mouth a chatter of nervous excitement.

Marvin smiled. 'Bert, it appears something strange has grown out of the top of your briefcase and it's making a strange beeping sound, as though it's going to explode.'

Bert rose from the sand, like a scissor lift. 'That must be an emergency message from Morris,' he whispered to Danny, as Marvin helped his Grandfather to his feet.

The Chief shook Bert's hand, but he held on to Danny's and

looked him deep in the eyes. 'The path to glory is rough and many gloomy hours obscure it, but remember the soul would have no rainbow if the eyes had no tears. Misfortunes do not flourish particularly in our path, they grow everywhere.'

Danny dropped his head at the reference to his looming premature death. But as the Chief continued, unaware that he'd been fooled, his words reminded Danny of his last visit to the hospital; his Dad's gaunt face smiling as usual, and his eyes brimming with excitement as he talked about the wonderful possibilities of life and how to live it, and Danny told the stories of his school day, too young to understand that his dad was saying goodbye.

'There is no Indian word for Religion. There is no fixed dogma or list of written rules, there is only an understanding that each must find their own path and live right with Nature and right with the world. It is your path and yours alone, and you will be forever known by the tracks you leave.

'I give you, my brother, the name *'Two Moons'*. In you, as in all men, are natural powers. You already possess everything necessary to become great and I see a new moon rising within you. Walk the good road my brother. Be strong with the warm strong heart of the earth. Walk on a rainbow trail; walk on a trail of song and all about you will be beauty. When you see a new footprint you do not know, follow it to the point of knowing, and when it comes time to die, sing your death song and die like a hero going home.

'There is no death only a change of worlds.'

22

Danny could never remember afterwards how he'd replied to the Chief – he wasn't sure he even said goodbye. He didn't remember Bert leaving either so it was a double surprise to be standing on the outside of the teepee with the old man charging towards him.

'Danny!' cried Bert, taking him by the arm and dragging him up to speed. 'We have to go.'

'What's wrong?'

'A big storm is heading for the airport. Where's Marvin?'

Marvin came bounding up and fell in next to Danny.

'Ah, Marvin. I have received some news.'

'Earl told me; they're going to close Flagstaff airport. That happens quite often with our summer storms. I have business to attend to before the helicopter is grounded and I'm afraid we will have to leave right away.'

'That suits us fine.'

'If my flight's cancelled then I don't have to go,' said Danny, as he was herded across camp, like a speed-walker caught up in a faster pack.

'They have brought forward the departure time to fly out as many people as possible,' said Bert.

'I don't want to fly out. Not yet.'

'I know, Danny.' Bert glanced around. 'But I think it is time we were leaving.'

Danny hadn't noticed that the camp was now completely deserted.

They emerged from the outer circle of teepees to find Earl flicking switches above his head, winding the helicopter into life.

'Why don't I miss this flight and stay another night with you?' shouted Danny, over the whine of the engines and throb of the accelerating rotor.

Bert helped Danny into his seat and pulled the lap belt across his stomach. 'It is better this way. I have never been good at saying goodbye.'

Finding a large Indian brave sitting on the other side of his rucksack, Danny turned, wide-eyed, back to Bert. 'You're not coming with me?'

'Marvin warned me there would not be room.' Bert shrugged and took hold of Danny's hand.

'This can't be it, Bert. What about your expedition?'

'Do not worry, my friend. I am sure we will meet again sometime in the future. Goodbye Danny and good luck.'

'I still don't understand my Jupiter trip,' shouted Danny, as the pitch of the engines increased. 'What about the accelerations I should have felt to change my frame relative to the Earth's?'

'I knew you would find the answer and there can be only one. According to the physics, your trip Danny –' A final roar from the engines forced Bert to close the door as he finished his sentence.

The helicopter lifted away leaving him small on the ground, and Danny with his hand pressed to the glass wondering if he'd said what he thought he'd said. He hadn't heard him so much as read his lips but he thought he said, '*Your trip, Danny, cannot have actually happened.*'

*

'I'll just set up your television screen,' said the flight attendant. 'The pre-flight video will be starting shortly.' She released the monitor from the wall panel next to Danny's seat and connected the headphones. 'Is there something else I can get for you? … No? Call if you need anything. The button's right here on your hand control.'

Danny took a slurp of ice-cold cola and slumped back in his armchair seat, staring at his dusty cowboy boots from beneath the brim of his Stetson.

He'd spent the helicopter journey watching the deepening blue sky and lengthening shadows, haunted by Bert's final words and failing to notice the lack of storm clouds gathering on any horizon. At Flagstaff he was met on the tarmac by an airline rep, piled into a baggage car, raced across the airfield and with nothing more than a glance at his rucksack, bundled up the steps of a plane, straight into first class.

The cabin lights dimmed and his screen burst into coloured light. But there was no flight attendant, eager to save his life in an emergency, just a vast brush-covered plain, a long straight road, and a tiny figure balancing on one leg. The camera panned down and right, revealing its position on top of a hill overlooking the plain. Three hundred feet below, the road swept around its base and an old VW Beetle, with a large and overloaded roof rack, reversed out of its shadow and onto the straight.

Danny sat up and put the headphones in place of his hat.

Down on the plain the Beetle crawled up to and then straight past the figure, and as the camera zoomed in on the face of a shocked and bewildered Danny, a shocked and bewildered Danny bolted to his feet and had a good look round the cabin.

A bald head rose above a copy of the New York Times behind him, and a woman's legs curled beneath a red dress across the aisle at the front. But as he couldn't see either of their screens, he dropped back to his own to watch himself running alongside the Beetle.

'Come on kid,' came a man's voice off-camera. 'Hop onto that running board.'

'I got twenty bucks ridin' on him,' said a second, younger man.

Then a woman. 'He'll do it,' she said, her voice clear, strong and right next to the microphone. 'You can tell he's that kind of kid. If for some reason he doesn't make it, the Old Man'll switch to Plan B ... but he'll make it.'

The camera stayed on Danny until his belly-flop onto the Beetle's bonnet, then panning left it refocused onto two men lying on their stomachs, cheering and doing a low-five with binoculars raised to their eyes.

'It's started,' said the woman, evidently rising to her feet before swinging the camera back onto the men. 'I'll see you boys back at HQ. Providing all goes to plan, we'll reopen the road in a little over an hour.'

Danny felt a prickle rise up the back of his neck as the picture cut to a small office and a woman perched on the edge of a desk. 'Hi

Danny,' she said, in the same earthy drawl, her blue eyes sparkling as she flicked a wisp of blond hair from her mouth. 'I'm Morris.'

Dressed in faded jeans and a yellow t-shirt with '*BUTAE is in the eye of the beholder*' rolling up and down across her chest, she was clearly female not male, American not French and only in her mid- to late-twenties.

'I expect you're mighty confused watching this. But then, as you've learned, reality isn't always as it appears. To begin with my name is Clare, but I represent Morris.' She checked her bracelet watch. 'It's 9:17 on Monday July 25th, and right now you're riding in an old VW Beetle, wondering what you've gotten yourself in to.' Clare slipped off the desk and her hand loomed large as it reached over the top of the camera and turned it towards the door. 'So let's see what's really going on.'

The door opened onto a corridor bustling with people, and they stepped into the current and fought their way upstream. 'It isn't normally this crazy,' said Clare from behind the camera, 'but everyone's caught-up in the excitement of the first day.'

The camera dived for a door marked *Production Office* – all box files and filing cabinets, and three busy women behind busy desks.

'Hi girls,' sang Clare. 'Say hello to Danny.'

'Hi Danny,' chorused the girls, pausing at their computers and telephones just long enough to smile and wave into the lens.

'Jenny, now everyone's at breakfast I'm takin' Danny on a quick tour, but call me if there's a problem. Thanks.'

The corridor had thinned and they continued down it, the camera swinging to glimpse in on the Production Manager cupped in his chair with his feet on his desk. 'No, it's a very specific flight plan,' he said into the phone, waving to the camera and rolling his eyes. 'They will be below 100 feet for ten minutes and only over the river.'

'We're having trouble securing permission for a special flight we have planned for you,' said Clare, as they passed the Art Department; a large, bright room filled with drafting tables, and model sections of Canyon wall complete with tiny bushes and trees. 'But, silly me, you'll know about that by now.'

The camera pushed through a set of double glass doors into the sunlight of a large car park bordered by three identically clad commercial buildings.

'We've been here three months setting everything up,' said Clare, swinging the camera to the left as she continued to cross the car park. 'Over there's our Props store, stacked high with teepee poles; piles of canvas, blankets and furs; shelves full of everything from clay pots to feathered bows; and duplicates of the crates on the Beetle's roof rack. The other half of the building houses our Costume Department, responsible for dressing the actors and 500 Native Americans for the powwow on the final day. None of this would be possible without the Havasupai and Hualapai Tribes. They own this area of the Grand Canyon and have given us permission to use their land. We've resurfaced twelve miles of public road and laid the section of new road leading right to the Canyon edge. We built the track and clearing into the scrub of the plain, the Beetle's rock platform at the top of the Canyon, and the track at the bottom that supposedly leads down to Rosie's, and when we're done it'll all be removed and re-landscaped.'

The picture panned back to their direction of travel and on to the second building; its industrial nature clear from the massive sliding door and yawning darkness of its interior. 'This is Special Effects,' said Clare, as the camera adjusted to the artificial light to reveal a spacious workshop. 'This is where they built the Beetle.'

In one corner pallets of sheet metal and tubular steel sat next to bench-lathes, presses and drills. A plastic-curtained paint shop stood in another corner, next to a hydraulic car ramp surrounded by electronic testing equipment and multi-drawer toolboxes on wheels.

The camera panned onto the middle of the workshop and settled on a maroon Citroen 2CV with broken windows, a concertinaed bonnet, and two heavy-duty trolleys doing the job of its front wheels.

'The Effects boys have prepared the Citroën for today's collision by firing it into a set of the Beetle's crash bars fixed to a concrete wall.' Clare laughed. 'Look at it now.'

She swept the camera onto the other side of the workshop and found the metal frame of a box, the size of a small house, with a telescopic-steel leg extended beneath each corner and a huge white ball trapped inside. The top and sides of the ball bulged through the open walls of its cage, but the bottom had been sliced off, level with the base, to leave a large open hole.

'We call that the Globe. The frame's rubber-coated and the ball's covered with the same alternating layers of super acoustic polymers used to soundproof the Beetle's cabin. The hydraulic legs are computer controlled so when … well, you'll see for yourself what it's for.'

A mobile started ringing and the picture cut to a door at the end of a short corridor, a white plaque bearing *MO.R.R.I.S* looming large in the screen.

Clare turned the camera onto herself, now dressed in white. 'It's Tuesday afternoon and I thought it was time you two met.' She placed her finger to the plaque and started moving it along the letters. 'Mobile Operations, Resources, Reconnaissance, Insertions and Supplies. Don't look at me; it was Ricky's idea. He's such a geek. This is my team and this is our control centre.'

The door opened onto a small, dark office, and Danny's eyes were drawn to three large monitors along the back wall. His own face searching the inside of the glass of the first screen; a Google results page for Albert Einstein on the second; and a picture of Bert under the heading *Einstein, the FBI Files* on the third.

'Yo, Clare,' said a young black guy, swivelling his chair through a three-sixty and returning to his keypad.

'Hi Ricky. Ricky's responsible for all the software controlling the Beetle. He's satellite-linked to its computer and the briefcase computer, and can relay information from the roadblocks and observation posts directly to the Old Man's earpiece.'

Ricky pushed a pair of thick, black-rimmed glasses up into a bush of curly black hair and pointed to the first screen. 'Hey, Danny my man. That's you bro!'

'You're online, searching for Einstein,' said Clare, 'but we are in control of everything you see and hear. We blocked the signal to your cell phone and reprogrammed its time. We printed the false newspapers, broadcast the radio excerpts and you are not really connected to the Internet. All of your requests are coming here to Ricky and he sends back pre-prepared pages doctored with photographs of Bert.'

'O-kaay,' said Ricky, his fingers a blur above the keypad. 'I think it's time to drop in our conspiracy theory.'

Danny's eyes popped wide on the monitor showing the picture from the briefcase web-cam, and as the cursor highlighted a Google entry on the monitor next to it, *The Einstein Mystery* appeared on the third.

'We'll leave you to it,' said Clare, backing the camera out of the control room.

'Ricky, how's he doing?' came a man's voice through the speakers – American with a trace of English but not a hint of German accent.

Ricky spoke into the microphone on his desk. 'Man, you were right again. He's hooked, reeled-in and 'bout to drop into the net.'

The door closed and the camera pushed through another set of glass doors into the car park, flaring briefly white in the late afternoon sun before Clare's face appeared on the screen.

'Yes, Danny, all this is revolving around you. But this is no film or TV show, though we are running it along the same lines and employing some specialists from that industry. Your experiences have shown that your reality is tied to your point of view, and this is the final lesson: a short video-diary of your science camp, but seen from *our* point of view.'

A swish of red fabric passed by in the aisle as a montage started unfolding on Danny's screen.

The Beetle driving along with its black window screens raised – filmed from an open-top jeep cruising alongside – suddenly nose-diving and disappearing backwards off the screen.

Four men in red overalls kneeling on either side of the road, and the screened Beetle rolling gently to a stop between them. The Citroen appearing from the other direction pushed by two more members of the crew; fat fingers of crumpled bonnet slipping between the Beetle's crash bars. Two men tightened the Citroen's grip with long-handled industrial pliers as two removed the trolley wheels from underneath its front, then all six ran towards a car-transporter and pickup parked fifty yards up the road.

Cheers resounded through Danny's headphones as the camera panned across a group in desert fatigues, slapping each other's backs in a dimly lit cave, and the red limestone wall behind them turned into a lattice of wooden scaffold and pink plaster. Brilliant-white light poured through a saucer-sized hole in the plaster wall, and expanding to fill the screen, it resolved into a little Beetle draped in parachute silk and three tiny figures watching a fourth jump up and down.

The picture bouncing as it approached the rear of the Beetle at the rear of a baggage train; two men carrying a blue cool-box, two a white metal chest, and an assortment of familiar looking crates in arms and on shoulders behind. As Bert appeared to talk to himself in the driver's seat, bodies crawled silently over the Beetle, like foraging ants, folding away two of the parachutes, setting up camp beneath the third, replacing every crate on the roof rack, and carrying the old cargo away.

Night-time, a Winnebago, a small six-wheeled truck, and a group of hooded ninjas replacing the white metal chest on its flatbed for one with IMAX taped to its side. They opened the original to reveal the fat cylinder of the telescope, and as the truck pulled away to the soft purr of an electric motor, a bald Bert appeared in the doorway of the Winnebago, watching it disappear into the dark with half his face peeling off.

Another pit stop, this time on a dirt track against a backdrop of grey shattered rock, with four telescopic-steel legs disappearing off the top of the screen to the distant throb of helicopter blades. The picture zoomed out to reveal the huge Globe in its frame and the Beetle rolling to a stop underneath. An array of four projectors rose out of the white chest on the roof rack, light already flashing through their lenses as the globe started swallowing the Beetle. The massive frame came to a stop, and as the crew chocked crossbeams against the Beetle's front and rear wheels – the only parts of her still visible – four cables started rising around the box frame. The camera followed the cables up to a single, thicker cable hanging from the middle of the helicopter's sticklike frame as it hovered high above. The cables straightened to form the outline of a towering pyramid, and the Globe lifted slowly off the ground and started to climb.

A telescopic leg descending towards the lens, a red-sleeved arm coming in from either side and grabbing hold of its rubber footplate as the picture darted between three other legs and three other members of the crew. The steel tubes pushed inside each other, like a shock absorber, and after the computer-controlled descent had lowered the Beetle to the ground, the arms dived for the crossbeam and shuffled it to the right. The cameraman's hands clipped a strop to the front of his harness, and a telescopic leg started extending beneath his dangling feet. With a gentle sway marking the transition to helicopter lift, they rose out over the Canyon, leaving the Beetle parked in a crumbling rock garage at its edge – and Danny grinning in his first class seat.

'It's Saturday,' erupted the Native American children, gathered around a woman in front of the Chief's teepee. 'Okay, off you go,' she said waving them away and immediately calling, 'Don't forget; no English once he arrives.'

'Hi Danny,' said Clare, her face filling the screen from between long black plaits and above a tasselled buckskin top. 'It's 1.00 pm on Wednesday July 27th, the final day, and it's a wonderful atmosphere here. You are due to arrive in 15 minutes then I'm going to zip off with Earl to make sure everything's ready at the airport. So bye for now, and see you very soon.'

23

The screen blinked off; the cabin lights on; and sensing immediately that he should, Danny turned to find his mum and a blotchy-faced bald man smiling over the top of his seat. 'Mum!'

'Oh, Danny,' she cried, pulling him to her. 'I'm so proud of you.'

Danny examined the bald man from over her shoulder. The stained moustache, liver spots and twenty-years-worth of sags and wrinkles were gone, but the nose and wiry eyebrows had been mostly his own, and though he looked more Mr Magoo than Einstein, Danny needed only one look into his pale blue eyes. 'Bert,' he laughed, peeling a blob of rubbery glue off his chin.

Bert dropped his jaw into a classic Einstein grin. Then without losing the sparkle in his eyes, his mouth closed, his eyes screwed tighter and a crease, running around the bottom of his face, joined his cheeks together like a Smiley. 'Hello Danny. I'm your Great Uncle Gilbert, but over here everyone calls me Bert.'

'Uncle Gil?'

'Yes,' laughed his mum, pushing him to arm's length, her eyes sparkling, her lips red, and her wavy, dark hair shining. 'He organised the whole thing.'

'I don't understand,' said Danny, looking back and forth between them.

Bert put his hand onto Danny's shoulder. 'You're father came to stay with me in New York in 1969, when he was only a year younger than you and I was a single man with absolutely no idea what to do with him.'

'You took him to the Kennedy Space Centre and to see the launch of Apollo 10,' said Danny.

'The whole country was gripped by the race to the Moon. It fired an interest in me and it had a big effect on your father.'

'He said that's why he became an engineer.'

'You were Alice's age the first time we met, but he told me about you over the years and after the funeral I kept in contact with your mother.'

'We'd speak every now and then,' said his mum, brushing the back of a forefinger under each eye and smiling at Uncle Gil. 'You always managed to get me to open up. Then last October, out of the blue, he phones and says we're coming to visit next summer and you're going to attend science camp. We met in London a week later and he told me all about it and asked if I'd agree.' She dragged Danny back over his seat. 'I so nearly didn't.'

'It was all for me?' said Danny, as his mum released him for some more damage-limitation on her makeup. He turned to Bert for an answer, concentrating on his eyes to find the man behind the mask.

'I designed everything to suit you,' said Bert, 'but it was to satisfy my own curiosity. Eleven years ago I bought the Missing Horse and retired here from New York, and for the first time in my life I saw the real night sky. Within a week I'd ordered a telescope and an interest in astronomy grew into a passion for cosmology. But like you, I'm no mathematician. The physicist Victor Weisskopf said that Einstein's theories made him feel like the peasant who asks the engineer how his steam engine works. The engineer explains how the steam goes through the engine, driving the piston up and down and turning the wheels, and the peasant replies, *Yes, I understand all that, but where's the horse?*'

Danny smiled. 'The Missing Horse.'

'I felt the same way. Book after book left my questions unanswered, or at least not explained in a way that worked for me. After groping my way to an understanding I realised that the best way to understand something is to experience it. When your mother told me you were flunking physics at school, I started putting together Bert's Universal Tour and Expedition.'

'So I've been a sort of physics guinea pig because I'm your average peasant?'

Bert laughed, but in a staccato chuckle rather than a full-blown guffaw. 'You have been far from average, Danny. And understood more than I could have hoped for.'

'It must've cost a fortune?'

'Yes, Gil,' said his mum, taking hold of Bert's arm. 'It must have.'

'I've been fortunate in life and money was not a problem; I would've spent twice the amount. Einstein once said that everyone who is seriously involved in the pursuit of science becomes convinced that a spirit or intelligence is revealed in the laws of the Universe, and I have become so convinced. But the Chief said it all. Everyone you have met was an actor working to our script, but though we arranged them, the Chief's words were not ours. They are the collected thoughts of his ancestors from Sitting Bull of the Sioux to Cochise of the Apache, recorded over the past two hundred years but containing the wisdom of thousands. *The voice of Nature holds treasures for us and understanding leads to appreciation. The day will come when we need to share those secrets; the time to start sharing is today.* There may be no place in our cities to hear the leaves of spring or the rustle of insect wings, but we can explore the wonder of Nature in our minds. That is the purpose of my expedition."

'And to help my grades?'

His mum looked down and squeezed his hand. 'I didn't agree to all this just to improve your grades,' she said, looking back up, the tears gathering again in her eyes. 'I agreed in the hope it would rekindle a lost interest and inspire you … like your dad used to.'

As Danny and his mum hugged each other tight, Bert slipped quietly away.

'Dad died like a hero, Mum.'

'Yes,' she spluttered, pulling him closer. 'He was the bravest hero of them all.' She pushed him away again so that he could see her smiling through the tears. 'You have been pretty brave yourself.'

'You should've heard me when we went off the top of Canyon,' said Danny, as they barged down the aisle with their arms around each other's waist. 'I thought I was going to die.'

'Don't,' she said, a shiver waving through her shoulders. 'I was a heartbeat away from withdrawing my consent.'

They passed the galley into an empty business class. 'Have you been here the whole time?' said Danny.

'Yes. I couldn't bring Alice in, she'd have given us away.'

'What are you all dressed up for?'

'We're going to a party.' She twirled away, her red summer dress swishing around her knees as she span on the soles of matching three-inch heels. 'How do I look?' she said, spinning to a stop with the hem of the skirt flared out in each hand and one knee bent behind the other.

'At the Missing Horse?'

'Thanks,' she laughed, putting her hands onto his shoulders and turning him back to face economy. 'We don't need a plane to get to the Missing Horse.'

She pulled aside the dividing curtain to a chorus of cheers, and as Alice ran forward Danny scooped her into his arms and grinned at the waves of smiling faces bobbing above the seats. There was Joe, Billy, little Christopher, the Indian boys from the powwow, a civilian and civil-looking Sheriff, Marvin dressed like a student, the old Chief dressed more like a cowboy, and crowded behind them, Clare, and the half-recognised faces of the office staff and crew from her video.

From the look of them, the party had already started.

*

'S' there y'are!' said Danny, jumping into his seat with a bottle of Apache in one hand and a pink balloon printed with $\mathbf{E = mc^2}$ in the other. He closed the door and watched the crowd of bodies silently gyrating on the dance floor, his mum's red dress spinning between the gaps.

The Beetle had been driven to Las Vegas that afternoon and parked, as guest of honour, in one corner of their large party room on the tenth floor of the Mirage Hotel. Danny was still trying to figure out how they'd got her up there.

'You're drinking Apache,' said Bert, holding up his own.

'Mum said tonight I can have h'as many h'as I like.' Danny took a swig, as though she might change her mind at any moment.

'That's why I drink it. It tastes good, but with no alcohol it keeps the mind sharp.'

Danny's mind was having too good a time to go sharp on him now and he grinned and took another lug of beer.

'It was fun, wasn't it?'

'Fhun?' Danny rocked his body off his seat and the balloon off his head. 'You beat me up. Pratikaly give me a heart attack.'

Bert laughed, nodding the dome of his head. 'And we drugged you.'

'You drugged me?'

'Just a mild herbal sleeping aid. We wanted you to sleep well so that you would be alert during the day, even with the excitement of flying into the Grand Canyon and losing three days in time.'

'I thought it was h'all the fresh h'air,' Danny told the balloon. He settled down to watch the revellers then bobbed back up. 'So you drugged me. Beat me up. Scare me half to death and make me think I'm going mad. I could shoo you.'

'That's nothing. Over the next two-and-a-half weeks at the Missing Horse, I'll show you what I've got planned for the next stage of our expedition.'

'H'it isn't h'over?'

'No,' said Bert, with all the enthusiasm of his alter ego, 'we're only a third of the way there. Einstein's theories describe spacetime and the large-scale operation of gravity in the macro-universe. But if we are to understand the development of the early universe in the intensely condensed conditions of the Big Bang, then we have to understand the structure and behaviour of the atoms from which everything is made. Einstein's theories don't work in this micro-universe, and we are now looking at sub-atomic particles to a scale as small compared to ourselves as the universe is large. At the beginning of the twentieth century another great branch of physics was developed to explain this behaviour: the theory of Quantum Mechanics.'

Danny was still grappling with the bit about scale.

'The quantum world is difficult to explain and almost impossible to understand. In fact quantum physics couldn't have been dreamt up it's so bizarre, yet it's the most successful theory we've ever discovered. Its influence has been huge on every branch of science from chemistry to microbiology and medicine to mechanics. It has led to the electronic and microelectronic revolution of this our modern age, furnishing us with everything from computers to 3-D printers and lasers to nanotechnology, and has passed every single

240

experiment that has ever been performed to test its validity. Einstein himself spent the latter half of his life trying to disprove quantum theory, and it is into this mind-blowing labyrinth that our expedition will venture next.'

Bert fell silent, staring out at the dance floor and contemplating the huge challenge ahead.

'Go on – then,' hiccupped Danny.

'Go on what?' said Bert, raising his bottle to his lips.

'Explain fontum quissics to me.'

Curling up, as though he'd been punched in the stomach, Bert sprayed his mouthful of beer at the windscreen.

The Journey

I hope you enjoyed Einstein's Beetle and feel that you gained an understanding of the physics.

Einstein once said that you do not really understand something unless you can explain it to your grandmother. Even if this is your first popular science book, you may find yourself trying to explain relativity to your friends and family, but with so much to remember and piece together from only one read through, don't be surprised if your explanation goes awry. I'd read scores of books before I tried, and my attempts led to the realisation that I needed to write this one.

If you do feel that way, then please help me extend the joy of understanding to others by posting a review on Amazon. You can also contact me directly on marksouthworth@einsteinsbeetle.com. If the book did not work for you, or you struggled with a particular aspect of the physics, I would be grateful if you would email me with your feedback so that I can improve the explanation for the future.

Einstein's Beetle is the first book in a cosmological trilogy that will explain our understanding of the universe right up to the present day. In Book II, *The Star of Africa*, which I am currently developing, Bert and Danny take off on a new adventure at CERN, the Centre for European Nuclear Research in Switzerland, to explore the micro world of the atom and the crazy world of Quantum Mechanics.

In the meantime check out the bibliography for further reading, and good luck with the rest of your journey.

About the Author

In 1986, having decided to quit training as a solicitor and travel the world, I jokingly informed my boss that I was going to become a stuntman. Whether it was destiny, dumb luck or the multiverse, at my first port of call in Western Australia I found an advert from a stunt team looking for new performers. For eighteen months I battled Zorro in live shows, and ninjas, drug dealers and zombies in 'C' movies, but then I kidnapped Kylie Minogue and was forced to return to the UK.

Since 1990 I have been a member of the British Register of Stunt/Action Coordinators and Performers, playing with the laws of physics on major Hollywood features. I have tried to kill Brad Pitt many times (*Troy, World War Z*) and taken beatings on his behalf (double in *Snatch*). Robert De Niro threw me forty feet into a statue (*Frankenstein*), Anne Hathaway dragged me down the side of an apartment block (*Batman, The Dark Knight Rises*), and I have been variously shot, stabbed, blown-up, knocked-down and set on fire;

and fallen from buildings, bridges, ships, cliffs and helicopters. Charlie Sheen tried to put his foot through my stomach (*The Three Musketeers*), Mel Gibson nearly removed my jaw (*Braveheart*), and I have broken my thumbs, two ribs, a collarbone and my back.

But I have been an astronaut floating in zero g (*Event Horizon*), a vampire ablaze in its coffin (*Interview with the Vampire*), a swashbuckling Pirate (*Pirates of the Caribbean, On Stranger Tides*), and jousting knight (*Covington Cross*). I have ridden beneath an elephant as Mowgli (*Jason Scott Lee double, The Jungle Book*), driven the silver Aston Martin DB5 as James Bond (*Pierce Brosnan double, The World is Not Enough* and *Tomorrow Never Dies*), snogged Mini Driver (*Mr Wroe's Virgins*) and punched Nigel Havers in the eye.

And, in the end, I did get to travel the world.

Quotations

Albert Einstein

All quotes are taken from CALAPRICE ALICE (Editor) THE NEW QUOTABLE EINSTEIN. © 2005 Princetown University Press and the Hebrew University of Jerusalem. Reprinted by permission of Princeton University Press.

p. III 'All physical theories, their mathematical expressions notwithstanding, ought to lend themselves to so simple a description that even a child could understand them' *Clark, Einstein, 344*

p. 93 'Teaching should be such that what is offered is perceived as a valuable gift and not as a hard duty.' *New York Times, October 5, 1952*

p. 194
'I made one mistake in my life – when I signed that letter to President Roosevelt advocating that the bomb should be built.' *To Linus Pauling, recorded in his diary*

'Had I known that the Germans would not succeed in producing an atomic bomb, I never would have lifted a finger.' *To Newsweek magazine, March 10, 1947*

p. 238 'Everyone who is seriously involved in the pursuit of science becomes convinced that a spirit or intelligence is clearly revealed in the laws of the Universe - a spirit vastly superior to that of man.' *To student Phylis Wright, 1936, Einstein Archive 42-601*

Native American Indian

Unless otherwise stated, all quotes are from:
NATIVE AMERICAN WISDOM by Kent Nerburn and Louise Mengelkoch © 1991. Reprinted with permission of New World Library, Novato, CA, USA.
INDIAN SPIRIT by Oren Fitzgerald © 2006. Reprinted with permission of World Wisdom Inc. Indiana, USA.
365 DAYS OF WALKING THE RED ROAD Copyright © 2003 by Terri Jean. Used by permission of Adams Media, an F+W Media, Inc. Co. USA. All rights reserved
THE SOUL WOULD HAVE NO RAINBOW IF THE EYES HAD NO TEARS by Guy A. Zona © 1994. Touchstone, Simon & Schuster Inc.

Permission sought from the Estate of Guy A Zona.

Marvin
p. 218

'Listen or your tongue will keep you deaf.' *Native American Proverb*, (Jean 2003)

'A pause giving time for thought is the truly courteous way of beginning and conducting a conversation.' *Chief Luther Standing Bear, Teton Sioux* (Nerburn and Mengelkoch 1992)

'It was chiefly owing to their deep contemplation in their silent retreats in the days of youth that the old Indian orators acquired the habit of carefully arranging their thoughts.' *Blackbird, Ottawa*, (Fitzgerald 2006)

Chief Shooting Star
p. 217

'With this sacred pipe you will walk upon the Earth … The bowl of the pipe is of red stone; it is the Earth. The stem of the pipe is of wood and this represents all that grows on the Earth … All the things of the Universe are joined to you who smoke the pipe.' *Black Elk, Oglala Lakota*, (Fitzgerald 2006)

'This tobacco comes from the whites …We mix it with the bark from the Indian trees and burn it together … So may our hearts and the hearts of the white men go out together and be made good and right.' *Blackfoot, Absaroke*, (Fitzgerald 2006)

'As we smoke the pipe and offer our prayer with each new day, we should remember the importance of having a sacred centre within us and that this sacred presence is represented by the pipe.' *Yellowtail, Absaroke*, (Fitzgerald 2006)

p. 218

'Friend and brother! It is the will of the Great Spirit that we should meet together this day. He orders all things and has given us a fine day for our council. He has taken his garment from before the sun and caused it to shine with brightness upon us.' *Red Jacket, Seneca*, (Fitzgerald 2006)

'We should understand well that all things are the work of the Great Spirit. We should know that he is within all things: the trees the grasses the rivers the mountains and all the four-legged animals, the winged peoples …' *Black Elk, Oglala Lakota*, (Fitzgerald, 2006)

'It is important to understand that there are many different ways of seeing

248

the world.' (Bruchac 1999)

'You must speak straight so that the words may go as sunlight to our hearts.' *Cochise, Chiricahua Apache Chief,* (Nerburn and Mengelkoch 1992)

p. 223
'The old people came literally to love the soil and they sat or reclined on the ground with the feeling of being close to a mothering power. It was good for the skin to touch the earth ...' *Chief Luther Standing Bear, Teton Sioux,* (Nerburn and Mengelkoch 1992)

'I will tell in my way how the Indian sees things. The white man has more words to tell you how they look to him, but it does not require many words to speak the truth.' *Rolling Thunder, Chief Josef, Nez Perce* (Nerburn and Mengelkoch 1992)

'I was born in nature's wide domain! The trees were all that sheltered my infant limbs, the blue heavens all that covered me. I am one of nature's children. I have always admired her. She shall be my glory; her features her robes and the wreath about her brow, the seasons, her stately oaks and the evergreen-her hair, ringlets over the earth - all contribute to my enduring love of her. And wherever I see her, emotions of pleasure roll in my breast, and swell and burst like waves on the shores of the oceans, in prayer and praise to the Great Spirit who has placed me in her hand.' *George Copway (Kahgegagahbowh), Ojibwe* (Nerburn and Mengelkoch 1992)

'When I was 10 years of age I looked at the land and the rivers, the sky above and the animals around me and could not fail to realize they were made by some great power.' *Brave Buffalo, Tatanka-Ohitika, Teton Sioux,* (Fitzgerald 2006)

'We believe that the Spirit pervades all creation ...' *Ohiyesa (Charles Eastman), Santee Sioux,* (Jean 2003)

'The Great Spirit is our Father but the Earth is our Mother.' *Bedagi, Wabanaki,* (Bruchac 1999)

'Every part of the Earth is sacred to my people, every shining pine needle every sandy shore, every mist in the dark woods, every meadow, every humming insect ...' *Chief Seattle, Duwamish- Suquamish,* (Jean 2003)

'The white man thinks with his head – the Indian thinks with his heart.' *Jimalee Burton, Cherokee,* (Jean 2003)

'The American Indian is of the soil ... He fits into the landscape, for the hand that fashioned the continent also fashioned the man for his surroundings. He once grew as naturally as the wild sun flowers; he belongs just as the buffalo belonged.' *Chief Luther Standing Bear, Oglala Lakota,* (Fitzgerald 2006)

'The old Lakota was wise. He knew that man's heart away from nature has become hard ...' *Chief Luther Standing Bear, Oglala Lakota* (Nerburn and Mengelkoch 1992)

'There is no quiet place in the white man's cities, no place to hear the leaves of spring or the rustle of insect wings.' *Chief Seattle, Suqwamish and Duwamish,* (Nerburn and Mengelkoch 1992)

'You red people will see the secrets of nature ... The day will come when you need to share the secrets with other people of the Earth because they will stray from their spiritual ways. The time to start sharing is today.' *Don Coyhis, Mohican,* (Jean 2003)

'Though I hear what the ground says ...' *Young Chief, Cayuse,* (Fitzgerald 2006)

'The planet itself calls to the other living species for relief ... The land waits for those who can discern their rhythms. The peculiar genius of each continent, each river valley, the rugged mountains, the placid lakes - all call for relief from the constant burden of exploitation.' *GOD IS RED: A NATIVE VIEW OF RELIGION by Vine Deloria Jr. © 2003. Reprinted with permission from Fulcrum Publishing, Golden, Colorado, USA.*

'How can the spirit of the earth like the white man ... everywhere the white man has touched it, it is sore.' *Wintu Woman,* (Jean 2003)

p. 224
'Then I was standing on the highest mountain of them all, and round about beneath me was the whole hoop of the world ... And I saw the sacred hoop of my people was one of the many hoops that made one circle, wide as daylight and as starlight ... But anywhere is the centre of the world ... Sometimes dreams are wiser than waking.' *Black Elk, Oglala Sioux,* (Jean 2003)

On the hoop of life there is a place for every species, every race, every tree and every plant. With all things and in all things we are relatives.

'With all things and in all things we are relatives.' *Sioux,* (Zona 1994)

'Being born as humans to this Earth is a very sacred trust. We have a sacred responsibility because of the special gift we have, which is beyond the fine gifts of the plant life, the fish, the woodlands, the birds and all the other living things on Earth. We are able to take care of them.' *Audrey Shenendoah, Onondaga,* (Bruchac 1999)

'Humankind has not woven the web of life. We are but one thread within it. Whatever we do to the web, we do to ourselves.' *Chief Seattle, Duwamish-Suquamish,* (Jean 2003)

'We are the physical mirroring of Miaheyyun, the total universe, upon this Earth, our mother.' *Fire Dog, Cheyenne,* (Jean 2003)

It is this completeness of life that must be respected in order to bring about the health of this planet.

'The ones that matter most are the children. They are the true human beings.' *Lakota,* (Zona 1994)

'We do not inherit the Earth from our ancestors, we borrow it from our children.' *Ancient Indian proverb,* (Jean 2003)

'When we walk upon Mother Earth, we always plant our feet carefully because we know the faces of our future generations are looking up at us from beneath the ground. We never forget them.' *Onondaga by Oren Lyons. From WISDOMKEEPERS: MEETING WITH NATIVE AMERICAN SPIRITUAL ELDERS by Harvey Arden and Steve Wall © 1990. Reprinted by permission of Harvey Arden and Steve Wall*

'That hand is not the colour of your hand ... The Great Spirit made us both.' *Luther Standing Bear, Oglala Sioux,* (Jean 2003)

'We are in this together my friends, the rich, the poor, the red, white, black, brown and yellow. We share responsibility for Mother Earth and those who live and breathe upon her.' *Leonard Peltier,* (Jean 2003)

'If you have one hundred people who live together, and if each one cares for the rest, there is one mind.' *Shining Arrows, Crow,* (Jean 2003)

'Good words do not last long unless they amount to something ... There has been too much talking by men who had no right to talk.' *Chief Josef, Nez Perce,* (Nerburn and Mengelkoch 1992)

'Your words circle like soaring birds which never land. I will try to catch them and take them back for my people to hear.' *Blue Jacket, Shawnee,* (Jean 2003)

'As we are going to part we will come and take you by the hand and hope the Great Spirit will protect you on your journeys and return you safely to your friends.' *Red Jacket, Seneca,* (Fitzgerald 2006)

'I will sing this song ... Songs are thoughts, sung out with the breath when people are moved by great forces and ordinary speech no longer suffices.' *Orpingalik, Netsilingmuit,* (Jean 2003)

p. 225
'... With beauty before me, I walk. With beauty behind me, I walk. With beauty below me, I walk. With beauty above me, I walk. With beauty all around me I walk. It is finished in beauty. It is finished in beauty.' *Ending to the House made of Dawn, from the Night Chant, Navajo,* (Bruchac 1999)

'The song is very short because we understand so much.' *Navajo,* (Zona 1994)

'Shadows are long and dark before me. I shall soon lie down to rise no more.' *Red Cloud, Oglala Lakota,* (Fitzgerald 2006)

'What is life? It is the flash of a firefly in the night. It is a breath of a buffalo in the winter time. It is as the little shadow that runs across the grass and loses itself in the sunset.' *Crowfoot, Blackfeet, 1880,* (Fitzgerald 2006)

'I shall vanish but the land over which I now roam shall remain and change not.' *Song of Hethushka Warrior Society, Omaha,* (Fitzgerald 2006)

'Oh that I could make that of my Red people and of my country, as great as the conceptions of my mind, when I think of the Spirit that rules the universe.' *Shooting Star, Tecumseh, Shawnee,* (Fitzgerald 2006)

'You look at me and you see only an ugly old man, but within I am filled with great beauty. I sit on a mountaintop and look into the future ... I see my people and your people living together.' *Old Man Buffalo Grass, Navajo,* (Bruchac 1999)

'The path to glory is rough and many gloomy hours obscure it.' *Black Hawk, Sauk,* (Nerburn and Mengelkoch 1992)

p. 226
'The soul would have no rainbow if the eyes had no tears.' *Minquass,* (Zona 1994)

'Misfortunes do not flourish particularly in our path, they grow everywhere.'

Big Elk, Omaha Chief, (Nerburn and Mengelkoch 1992*)*

'In all Native American languages still spoken today, there is no word for Religion ... There is no fixed dogma or a list of written rules, there is only an understanding that one is to seek one's own path and live right with nature and right with the world.' *Terri Jean,* (Jean 2003)

'Do not allow others to make your path for you. It is your road and yours alone. Others may walk it with you, but no one can walk it for you.' *Shooting Star, Tecumseh, Shawnee,* (Jean 2003)

'We will be forever known by the tracks we leave.' *Dakota,* (Zona 1994)

'In you, as in all men, are natural powers ... You already possess everything necessary to become great.' *Legendary Dwarf Chief, Crow,* (Jean 2003)

'Walk the good road ... Be strong with the warm strong heart of the earth.' *Anon, Sioux,* (Jean 2003)

'Walk on a rainbow trail; walk on a trail of song and all about you will be beauty.' *Navajo Song,* (Jean 2003)

'When you see a new trail or a footprint you do not know, follow it to the point of knowing.' *Uncheedah, Santee Sioux,* (Jean 2003)

'When it comes time to die ... Sing your death song and die like a hero going home.' *Aupumut, Mohican,* (Fitzgerald 2006)

'There is no death only a change of Worlds.' *Chief Seattle, Suqwamish and Duwamish,* (Nerburn and Mengelkoch 1992*)*

Bibliography

Barrow, John D. (1994) *The Origin of the Universe*. Basic Books.

Bernstein, Jeremy (1997) *Albert Einstein and the Frontiers of Physics*. Oxford University Press.

Bodanis, David (2001) $E=mc^2$. Berkley Trade.

Bragg, Melvyn (2007) *On Giants' Shoulders*. John Wiley & Sons, Ltd.

Breithaupt, Jim (2000) *Einstein: A Beginner's Guide*. Hodder & Stoughton.

Bruce, Colin (1997) *The Strange Case of Mrs Hudson's Cat*. Perseus.

Bruchac, Joseph (1999) *Native Wisdom*. HarperSanFrancisco.

Calaprice, Alice (2000) *The Expanded Quotable Einstein*. Princeton University Press.

Clegg, Brian (2007) *Light Years*. Palgrave Macmillan.

Cole, K.C. (1999) *The Universe and the Teacup*. Harcourt Brace International.

Coleman, James A. (1990) *Relativity for the Layman*. Penguin.

Coles, Peter (2000) *Einstein and the Birth of Big Science*. Totem.

Coveney, Peter and Highfield, Roger (1990) *The Arrow of Time*. W.H. Allen.

Davies, Paul (1989) *Superforce*. Unwin.

Davies, Paul (1993) *The Mind of God: Science and the Search for Ultimate Meaning*. Penguin.

Davies, Paul (1996) *About Time: Einstein's Unfinished Revolution*. Penguin.

Einstein, Albert (1996) *Ideas and Opinions*. Crown.

Ferguson, Kitty (1999) *Measuring the Universe*. Headline.

Feynman, Richard P. (1990) *QED: The Strange Theory of Light and Matter*. Penguin.

Feynman, Richard P. (1998) *The Meaning of It All*. Penguin.

Fitzgerald, Michael Oren (2006) *Indian Spirit*. World Wisdom Books.

Gamow, George (1993) *Mr Tompkins in Paperback*. Cambridge University Press.

Greene, Brian (2005) *The Elegant Universe*. Vintage.

Gott, J Richard (2002) *Time Travel in Einstein's Universe*. Phoenix.

Gribbin, John (1996) *Companion to the Cosmos*. Phoenix.

Gribbin, John (1996) *Schrodinger's Kittens and the Search for Reality.* Phoenix.

Gribbin, John (1999) *Almost Everyone's Guide to Science.* Phoenix.

Gribbin, John and Gribbin, Mary (1997) *Newton in 90 Minutes.* Constable.

Guth, Alan H. (1998) *The Inflationary Universe.* Vintage.

Hawking, Stephen (1994) *Black Holes and Baby Universes and Other Essays.* Bantam.

Hawking, Stephen (1995) *A Brief History of Time.* Bantam.

Highfield, Roger (2002) *Can Reindeer Fly?* Phoenix.

Jean, Terri (2003) *365 Days of Walking the Red Road.* Adams Media Corp.

Kaku, Michio (1994) *Hyperspace.* Oxford University Press.

Kolb, Rocky (1996) *Blind Watchers of the Sky.* Perseus.

Krauss, Lawrence (2007a) *Fear of Physics.* Basic Books.

Krauss, Lawrence (2007b) *The Physics of Star Trek.* Basic Books.

Moore, Patrick (1999) *Atlas of the Universe.* Philip's.

Moring, Gary F. (2002) *The Complete Idiot's Guide to Einstein.* Alpha Books.

Nerburn, Kent and Mengelkoch, Louise (1992) *Native American Wisdom.* New World Library.

Pirani, Felix and Roche, Christine (2006) *Introducing the Universe.* Icon Books.

Schwartz, Joseph and McGuinness, Michael (2005) *Introducing Einstein.* Icon Books.

Smoot, George (1995) *Wrinkles in Time.* Abacus.

Sobel, Dava (2000) *Galileo's Daughter.* Fourth Estate.

Stannard, Russell (2005) *The Time and Space of Uncle Albert.* Faber & Faber.

Strathern, Paul (1997) *Einstein and Relativity.* Arrow.

Strathern, Paul (1998) *Newton and Gravity.* Anchor.

Waugh, Alexander (1999) *Time From Micro-seconds to Millenia - the Search for the Right Time.* Headline.

Weinberg, Stephen (2000) *Dreams of a Final Theory.* Orion.

White, Michael (1998) *Isaac Newton, The Last Sorcerer.* Fourth Estate.

White, Michael and Gribbin, John (1997) *Einstein, A Life in Science.* Pocket Books.

Zimmerman, Barry E. and Zimmerman, David J. (1995) *Why Nothing Can Travel Faster than Light.* Cassell.

Zona, Guy A. (1994) *The Soul Would Have No Rainbow If the Eyes Had No Tears.* Touchstone.

Printed in Great Britain
by Amazon